T0305879

Pavement Drainage

Theory and Practice

Pavement Drainage

Theory and Practice

G.L. Sivakumar Babu
Prithvi S. Kandhal
Nivedya Mandankara Kottayi
Rajib Basu Mallick
Amirthalingam Veeraragavan

CRC Press
Taylor & Francis Group
Boca Raton London New York

CRC Press is an imprint of the
Taylor & Francis Group, an **informa** business

CRC Press
Taylor & Francis Group
6000 Broken Sound Parkway NW, Suite 300
Boca Raton, FL 33487-2742

ISBN 13: 978-0-8153-5360-7 (hbk)

Visit the Taylor & Francis Web site at
www.taylorandfrancis.com

and the CRC Press Web site at
www.crcpress.com

Dedication

In loving memories of Professor Braj Bhushan Pandey (1940–2018), Former Professor and Department Head, Civil Engineering, Indian Institute of Technology (IIT) Kharagpur, and Professor C. E. G. Justo (1935–2019), Former Professor and Head, Department of Civil Engineering and Principal, UVCE., Bangalore University, Bangalore, who dedicated their lives toward pavement engineering and inspired generations.

Contents

Preface

This book is offered to students and practitioners of Civil and Environmental Engineering as a textbook and also as a reference book on pavement drainage. It is expected that they will find it useful for understanding the sources of water in pavements, concepts and principles of design and construction of drainage structures, and information about state-of-the-art practices in pavement drainage.

Except in very dry regions of the earth, one can always expect some water in pavements – roads, highways, parking lots, airfields and container terminals. The water may or may not be visible on the surface, and may be in one or more of the underlying layers. As a result, the pavement may develop surface and subsurface drainage problems, and deteriorate prematurely – leading to distresses in the pavement, loss of serviceability, unsafe riding conditions, endless cycles of repairs and maintenance, and wastage of money to the agency and the road user, time and effort and higher life cycle cost.

Where does the water come from? It can come from rain (or in cold regions, melted snow), subsurface sources as high water table and aquifers, or water that has been trapped in one or more of the pavement layers. Whatever may be the source, the presence of water in a pavement system is not desirable for the proper functioning of a pavement.

Water (as liquid water or moisture) creates problems in each and every layer of pavements, and the severity and extent of the distresses depend upon the quantity and the state of the water. For example, static water in granular base layers can create a problem through a reduction of the structural strength of the entire pavement, whereas flowing water can dissolve carbonate materials present in any of the underlying layers and can result in sinkholes and complete collapse of a large part of a pavement.

Water related problems can, and do occur, in all types of pavements. For example, in thin asphalt surfaced pavements, water can infiltrate easily through the mainline or shoulder areas and deteriorate granular layers; in thick asphalt surfaced pavements water enters through cracks in the surface; in pavements with stabilized bases, shrinkage cracks lead to reflective cracking and ingress of water; and in concrete pavements water enters through joints and cracks and lead to pumping, erosion of foundation material, and reduction in support.

Even though a significant amount of progress has been made over the past few decades in materials, design and construction practices, damage and distresses caused by water in pavements still remain the number one concern in many parts of the world. At the same time, in most places, the population keeps growing, vehicle and aircraft usage keep increasing, and hence the demands on the pavements keep increasing. On the other hand, precipitation is projected to increase, as a result of climate change, in many areas of the world. This means that the volume of water, and the time duration to which the pavements are exposed to water are both expected to increase in the future. How can we make pavements resilient to the impacts of more drainage related issues in the future?

There are also concerns about sustainability. For economic reasons pavements, which require a large quantity of materials (for example, >20,000 tonnes of good quality aggregates per lane-km of roadway) are mostly made up of naturally occurring aggregates. Cement, used in concrete (or "rigid") pavements is manufactured from limestone and asphalt binder, used in asphalt (or "flexible") pavements is processed from crude oil. There are two very serious concerns that are facing the pavement community. First, we are running out of good quality construction aggregates. Second, we realize that production of cement and asphalt binder, and the associated mixes result in generation of high amounts of CO_2 (the primary cause of global warming), emissions of other undesirable gases (such as SO_2, NO_x), and expense of considerable amounts of energy – energy that comes chiefly from fossil fuel. If we keep on losing our pavements to water induced damages and need to rebuild them quite frequently, how are we going to afford the materials and energy, and prevent the generation of harmful emissions?

To address sustainability, many agencies are resorting to in-place recycling such as cold in-place recycling (CIR) or Full Depth Reclamation (FDR) with relatively thin surface layers in flexible pavements. While these technologies provide very cost effective solutions for rehabilitation of pavements, there is a growing body of literature that shows failures of these layers within very short periods of time, primarily due to the impact of moisture. As these layers consist of unbound materials, they cannot resist the effect of moisture, their hydraulic conductivity properties are unknown, and generally these layers are not drained. One wonders how is the water that infiltrates through the surface channeled effectively out of the layer? And if it is not, why should a relatively costly reclaimed layer be provided that will deteriorate within a short period of time? It defeats the whole purpose of using a sustainable pavement construction technology.

Summing up, we can conclude that the duration of time pavements will be exposed to water, and the volume of water are projected to increase and the demands for pavements keep increasing, while resources for building pavements are dwindling. Hence, in the authors' opinion, it is very critical to revisit the impact of water in pavements at this time, understand the impacts, and bring out available and potential solutions.

One more thing. The level of technological development for the modification of the naturally occurring aggregates, through the introduction of specifications on shape, size and gradation has most likely reached a plateau. Although significant advances have been made in developing highly engineered products such as additives (for example, polymers), the extent to which they are used in pavements is still very small, and pavements are still made up of 90%–95% natural aggregates. And to keep the costs within budget, it will be the same, for the foreseeable future. This means that we have practically reached a plateau as far as the improvement of individual layers of pavements is concerned. The authors' opinion is that under such a circumstance, we need to pay more attention to the pavement structure as a whole – and develop the most cost-effective "pavement system" that is capable of draining and resisting the impact of water.

This book, therefore, as much as it is possible, focuses on the pavement as a whole and not only on the individual layers. It also tries to keep the connection between the three essential components of the pavement system – hydrology, hydraulics and structures – alive and prominent in most of the discussions.

This book draws from a number of important sources as mentioned in the reference list at the end of each chapter. Readers are requested to check the resources available at the Highway Drainage website of the US Federal Highway Administration(FHWA)(www.fhwa.dot.gov/engineering/hydraulics/highwaydrain/). Finally, due to the growing emphasis on connecting infrastructure design with climate change, readers are advised to check the recommendations of US National Cooperative Highway Research Program (NCHRP) project: 15–61 Applying Climate Change Information to Hydrologic and Hydraulic Design of Transportation Infrastructure (https://apps.trb.org/cmsfeed/TRBNetProjectDisplay.asp?ProjectID=4046); the report is not available at the time of writing of this book.

The authors will be grateful for any feedback from readers to improve the book in the future.

Please visit the book's CRC Press webpage at https://www.crcpress.com/p/book/9780815353607 for downloadable PowerPoint slides which include quizzes, schematics, figures, and tables.

Acknowledgments

Students, colleagues, friends and family members have made this book possible, through their untiring help, contributions and encouragement. Mr. Shubham Kalore, PhD student, Indian Institute of Science (IISc), Bangalore, India, and Mr. Asadi Mojtaba, University of Texas, El Paso have provided results of their original research for this book. Col. Shailendra Grover, former M.Tech student of IIT Madras contributed a significant part of his literature review. Mr. Mathias (Matt) Collins of the National Oceanic and Atmospheric Administration (NOAA), and Mr. Wade McClay, Mr. Rick Bradbury and Mr. Dale Peabody of Maine Department of Transportation provided much needed information and figures.

Special thanks go to Professor Soheil Nazarian, University of Texas, El Paso, for his help and encouragement. We thank Professor Krishna Prapoorna Biligiri, Indian Institute of Technology, Tirupati, Dr. Ellithy S. Ghada, US Army Corps of Engineers, Mr. David Hein, P.E./P.Eng., Applied Research Associates, Inc,. Prof. Dr.-Ing. Marc Illgen, University of Kaiserslautern, Minnesota Department of Transportation and Professor John C. Stormont of the University of New Mexico, Dr. Prajwol Tamrakar of Tensar International Corporation and Dr. Pavana Vennapusa, Ingios Geotechnics, Inc. for granting permissions for very valuable figures.

We are grateful for the encouragement from our family members. Special thanks go to Uma Kandhal, Arun Prakash Akkinepally, Sumita and Urmila Basu Mallick.

Finally, this book was made possible with the very helpful staff of Taylor & Francis Group, especially Mr. Joe Clements and Ms. Lisa Wilford. It has been a pleasure to work with them.

Authors

Prof. G.L. Sivakumar Babu is a professor of Civil Engineering, and also serves as Chairman, Centre for Continuing Education, Indian Institute of Science, Bangalore. He worked in Central Road Research Institute, Airports Authority of India, was a member of the Highway Research Board of Indian Roads Congress, and is associated with the development of codes for pavement design. He works extensively on reliability and risk assessment of civil engineering systems, ground improvement, geosynthetics and sustainability. He is currently President of the Indian Geotechnical Society. He is also the Governor for ASCE Region 10. He has contributed significantly internationally, and published approximately 250 papers in the area. He is the Chairman of the Technical Committee on Forensic Geotechnical Engineering of the International Society of Soil Mechanics and Geotechnical Engineering. He is the recipient of many awards and distinctions nationally and internationally.

Prof. Prithvi S. Kandhal is Associate Director Emeritus of the National Center for Asphalt Technology (NCAT) based at Auburn University, Alabama. NCAT is the largest asphalt road technology center in the world. Prior to joining NCAT in 1988, he served as Chief Asphalt Road Engineer of the Pennsylvania Department of Transportation for 17 years. He has held three prestigious national and international positions: President, Association of Asphalt Paving Technologists (with members from around the world); Chairman, American Society for Testing and Materials (ASTM) International Committee on Road Paving Standards (responsible for more than 200 standards used worldwide); and Chairman, US Transportation Research Board Committee on Asphalt Roads. Prof. Kandhal has published more than 120 technical papers and has coauthored the first ever textbook on asphalt road technology. In 2011, he was inducted on the "Wall of Honor" established at the largest asphalt road research center in the United States. In April 2012, he was made honorary member of the International Association of Asphalt Paving Technologists.

Dr. Nivedya Mandankara Kottayi held a post-doctoral position in the Department of Civil and Environmental Engineering, Worcester Polytechnic Institute (WPI), Massachusetts. Prior to joining WPI, she held both a post-doctoral research fellowship and a graduate research assistantship at Indian Institute of Technology Madras, Chennai, India. She works on the mechanical characterization of bitumen stabilized mixtures which consist of reclaimed asphalt material and foamed asphalt. Since joining WPI she has expanded her research to include moisture damage of asphalt mixtures, drainage characteristics and finite element analysis of pavements under extreme climatic conditions, nanoscale characterization of asphalt with antistripping additives, and application of artificial intelligence in pavement engineering.

Dr. Rajib Basu Mallick is a graduate of Jadavpur University, India, and has worked as a research assistant and as a senior research associate at the National Center for Asphalt Technology, Auburn University. He is currently the Ralph White Family Distinguished Professor of Civil and Environmental Engineering at Worcester

Polytechnic Institute (WPI). He has coauthored more than 100 papers for journals and conference proceedings, as well as several practical reports, manuals, and state-of-the-practice reports for federal, state and local highway agencies. He has taught professional courses on asphalt technology, presented courses on recycling for the FHWA, as well as lectured in national workshops at the Indian Institute of Technology (IIT) at Kanpur, Kharagpur, and Madras in India. He is a member of several professional organizations and a registered professional engineer (PE) in Massachusetts. He has served as a consultant on several projects for practitioner organizations, worked at the University of Peradeniya, Sri Lanka as a Fulbright Fellow and has served IIT Madras as a professor of civil engineering for one year.

Dr. Amirthalingam Veeraragavan is a Professor in the Department of Civil Engineering at IIT Madras. He has more than 38 years of experience in teaching, research and industrial consultancy. He received several awards from the Indian Roads Congress for his best and highly commended research papers. He is a recipient of the UGC Career Award for young teacher and National Award for Promising Engineering College Teacher in India, as well as the Viswakarma Award for Outstanding Academician/Technologist/Researcher/Innovator from the Construction Industry Development Council. He has published more than 100 research papers in journals and 150 papers in international conference proceedings. He is an author of books on surveying, highway engineering and highway materials and pavement testing. He is a Fellow of the Institution of Engineers (India) and a Fellow of the International Society for Engineering Asset Management. He is a member of the Highway Research Board, IRC Committees on Specifications and Standards, Flexible Pavements, Composite Pavements, and Road Maintenance and Asset Management.

1 Pavements – Overview

1.1 WHAT IS A PAVEMENT AND WHY DO WE NEED IT?

A pavement (Figure 1.1) is an engineered structure that provides a smooth, safe and stable platform for the movement of vehicles and pedestrians. A pavement is needed since in most cases the existing soil layer is not capable of supporting the loads of commonly used vehicles. Without the pavement the soil layer would deform excessively and hinder the movement of vehicular traffic with a desirable speed and safety.

Pavements can be primarily classified into four types:

1. Flexible
2. Rigid
3. Composite
4. Unsurfaced

Flexible pavements consist of the soil subgrade, uncrushed or crushed aggregate subbase and base, and binder and surface layers with asphalt (bitumen) aggregate mix – Hot Mix Asphalt (HMA) (predominantly) or cold mix asphalt.

Rigid pavements are made up of soil subgrade, subbase and base as in the case of flexible pavements, and Portland Cement Concrete (PCC) surface layer. Generally, the PCC layer is without any reinforcement, and has transverse joints (Jointed Plain Concrete Pavement, JPCP), but there can be Jointed Reinforced Concrete Pavement (JRCP) as well as Continuously Reinforced Concrete Pavements (CRCP). Thin (150 mm) and ultrathin (<150 mm) concrete layers called "whitetopping" and Precast/Prestressed Concrete Panels (PCP) are also becoming popular for rehabilitation of existing pavements.

Composite pavements consist of two or more layers of materials that use different types of binders – for example HMA and PCC layers in the same pavement (Smith, 1963). This may be HMA layer(s) over PCC layers or white-topping (PCC) layers over existing HMA layers. The HMA layers are provided to improve the ride quality of existing PCC layers, whereas thin and ultrathin PCC layers are now being constructed for rehabilitation of existing HMA layers.

Unsurfaced pavements are generally made up of soil and aggregate layers, without any bound (asphalt aggregate or PCC) layer on the top. They are generally stabilized

FIGURE 1.1 Asphalt (left) and concrete (right) pavements.

with additives (dust palliatives) for reducing the moisture susceptibility and dust potential of the surface layers. The use of heavily polymer modified soil layers as unsurfaced roads is increasing.

In addition, there are various types of pavements that are constructed for specific uses such as interlocking concrete panels for permeable pavements and porous asphalt pavements for rainwater harvesting, and flexible pavements that are grouted with cement slurry for heavy loading pavements.

1.2 LAYERS AND THEIR FUNCTIONS

Generally, a pavement consists of a number of layers (courses), each of which serves a specific function. In a flexible pavement, the soil subgrade serves as a stable platform for building the rest of the structure, while the subbase and base serve as platforms for constructing the overlaying HMA layers, as well as for distributing the load over a larger area so as to reduce the stresses in the subgrade to an acceptable level. In addition, if designed for that purpose, the base layer may function as a drainage layer also, to channel the water entering the pavement structure from above or below to side ditches or an edgedrain system which is installed within the subbase or subgrade. The HMA binder layer serves as a load distributing layer, while the surface layer acts primarily as a riding or wearing layer that protects the lower layers from the ingress of air and water. The pavement structure is designed in a way such that the repeated tensile stresses/strains resulting from repeated traffic loading are low enough so as not to cause cracking failure before the end of the design life, and such that the compressive stresses/strains are low enough so as not to cause rutting (permanent deformation) failure of the pavement before the end of the design life.

A relatively new concept that has been developed is that of a perpetual pavement (Newcomb et al., 2001), in which the structure is designed in such a way that neither the tensile nor the compressive stresses exceeds a maximum threshhold value (endurance limit), such that the pavement does not ever fail – it survives for perpetuity. The only deterioration will be at the wearing layer that is the surface and the shallow layers, which can be maintained and rehabilitated at regular intervals.

In a rigid pavement, the stiffness of the PCC layer is generally much higher than the lower layers, and hence it serves as the primary load bearing structure through

FIGURE 1.2 Layers in pavements.

slab action. The functions of the bottom layers (Figure 1.2) are generally drainage (base layer), stability and durability (subbase and subgrade). In this case, the lower layers serve as a combined supporting platform for the PCC layer. In JPCP (which are the most widely used type, ACPA, 1999) dowels are provided at joints for load transfer between successive slabs, and in JRCP and CRCP the reinforcements are provided for supporting of thermal stresses and vehicular loads in the pavement structure.

An example of an unconventional pavement is the inverted pavement or stone interlayer pavement system. In this case, the successive layers (from the top) consists of a relatively thin HMA layer, an unbound aggregate base (UAB), a cement treated base (CTB) or a subbase over the subgrade. The UAB in this case can be compacted to a much higher density than that obtained in a conventional pavement structure, because of the presence of the stiff CTB (which acts as an "anvil" during compaction). Because of this and the much higher stress (from traffic loading) due its proximity to the surface, a relatively high resilient modulus can be achieved in the UAB in the inverted pavement. This advantage is utilized to cut down the cost of providing a thicker HMA layer on the surface.

1.3 RELEVANT PROPERTIES OF MATERIALS

The structural design (thickness of different layers) of pavements is generally carried out on the basis of either empirical or Mechanistic-Empirical (ME, increasing use) methods. The primary basis of this design is the consideration of the relevant material or layer properties. Generally, all of the layers in both flexible and rigid pavements are designed as elastic layers – primarily because of the simplicity of the approach and the relatively low stresses that are generated, specifically in the lower layers. Elastic modulus is the key parameter in this approach and to simulate representative loading, in the laboratory the modulus is determined by different methods for the different layers. For soil and aggregate layers, the Resilient Modulus (M_r) is used, whereas for HMA and PCC layers, the Dynamic Modulus (E*) and Modulus of Elasticity (E) are utilized, respectively. In reality, both soil and aggregate materials may exhibit characteristics of nonlinearity as they are stress dependent materials, whereas asphalt mixes tend to exhibit features of viscoelastic materials. To take these factors into account, the material specific tests are conducted and appropriate material properties and/or constitutive models are considered in the design. For example, stress and saturation dependent models of soil and aggregate layers have been developed and utilized in ME design. Furthermore, soil layers can also contain frozen and thawed water during some parts of the year. To characterize the soil and aggregate layers (unbound layers) effectively, one needs other parameters such as specific gravity, optimum moisture content, maximum dry unit weight and hydraulic conductivity, as well as dry heat capacity and thermal conductivity. The Soil Water Characteristic Curve (SWCC) needs to be estimated.

In the case of HMA, whose behavior is highly temperature and loading-time dependent, a full characterization of the material over a range of expected temperatures and loading-time is utilized. Furthermore, in addition to E*, Flow Number (FN) is used for high temperature, and creep compliance (C_c) and indirect tensile strength (ITS) are used for low temperature characterization. In addition, other properties such as thermal conductivity, heat capacity and surface shortwave absorptivity are also used.

Note that a proper ME method of designing flexible pavements requires a significantly larger set of test properties than what is required for empirical design. To facilitate the use of ME design, in the absence of availability of appropriate test equipment, equations relating simple test results to more appropriate test properties have been developed. For example, one may use California Bearing Ratio (CBR) of a soil or an aggregate to estimate its resilient modulus through empirical relations. However, such use should be made with caution – as the conditions for the simpler tests are neither appropriate nor simulative of actual loading conditions. A more appropriate method is the use of statistical equations that have been developed to predict mechanistic properties on the basis of volumetric properties (which can be determined relatively easily) and the properties of the individual components.

For PCC layers, the relevant properties consist of gradation, compressive strength, elastic modulus and coefficient of thermal expansion. The structural design is based on the concept of limiting stresses and deformations to prevent excessive damage and deterioration of pavements. Overstressed pavements due to traffic loads and environmental effects will result in pavement distresses such as fatigue cracking, faulting, pumping, punchouts, and curling and warping. For analysis of stresses, a concrete pavement is idealized as a rigid slab resting on a spring-like foundation. Theories are used to develop equations relating stresses and deformations to material and structural properties, and then analytical methods are used to predict these stresses and deformation on the basis of properties such as modulus, Poisson's ratio, and stiffness of the subgrade, as well as temperature fluctuations. The design process essentially consists of consideration of traffic and environmental factors and the use of the relevant properties (through material characterization tests) to ensure that the slab can sustain the stresses for the duration of its life, without failing. The free flowing of subsurface water is critically important to durability, and resistance to pumping, and frost action in PCC pavements. Bases may be bound or unbound. A granular base or a cement- or asphalt-treated base may be used. If bound material is used, an open-graded mixture should be used to provide adequate drainage and water flow.

Climate and weather, and drainage related parameters represent a very critical set of inputs in structural design of pavements. In the Enhanced Integrated Climate Model (EICM) that is incorporated in the AASHTO (American Association of State Highway and Transportation Officials, www.transportation.org/) ME design method (currently known as the AASHTOWare Pavement ME Design, http://me design. com/MEDesign/?AspxAutoDetectCookieSupport=1), hourly air temperatures, wind speed, precipitation, percent sunshine and relative humidity are utilized. The groundwater table depth at various seasons, which affects the moisture content of soil, aggregate layers and hence their moduli, is required. For drainage properties, infiltration (as percent of precipitation), the presence or absence of drainage layers within the pavement structure, and that of edgedrains, are needed. Note that the amount of water entering the unbound layers, and their effects are currently considered – the bound layers are assumed to be unaffected by water. In the original AASHTO method of pavement structural design, drainage coefficients were used along with structural coefficients for flexible pavement courses to calculate the Structural Number (SN).

Table 1.1 shows a summary of properties that are needed for different types of pavement materials for structural design.

1.4 FACTORS THAT INFLUENCE THE LIFE OF PAVEMENTS

Load and climate-related factors – precipitation and temperature, and the effect of temperature on water in the pavement, are the principal causes of deterioration of pavements over time. The structure of the pavement is designed on the basis of the material and layer properties. Generally, the material properties of the pavement layers are determined from tests that are conducted under representative conditions – conditions that mimic expected conditions in the field over the entire design life of the pavement. The success of the design work is dependent on, to a large extent, how close the laboratory testing conditions are to the expected field conditions, and to what extent the ancillary structures are provided to make the field conditions similar to the "expected" conditions.

One relevant example in this regard is the provision of drainage. If the soil and aggregate layers are expected to be at a certain moisture content/saturation for a certain period of time, and the material properties tested under those conditions are used in the design, then the drainage structure of the pavement should be provided in such a way that in the field the soil and the aggregate layers are indeed exposed to the expected moisture content/saturation levels for the expected period of time. In most cases, it is a disparity between laboratory testing and actual field conditions that are responsible for premature failure of pavements.

The presence of water reduces the structural capacity of all layers that are used in conventional pavement, in a significant way, except for PCC layers. However, PCC

TABLE 1.1
Inputs for Design of Pavements

Materials Type	Inputs Required		
	Required for Response	**Required for Distress**	**Required for Climate Models**
HMA	Dynamic modulus of HMA, Poisson's ratio	Tensile strength, creep compliance, coefficient of thermal expansion	Surface shortwave absorptivity, thermal conductivity and heat capacity, asphalt binder viscosity (stiffness) characteristics
PCC	Modulus of elasticity, Poisson's ratio, unit weight, coefficient of thermal expansion	Modulus of rupture, split tensile strength, compressive strength, cement type, cement content, water to cement ratio, ultimate shrinkage, amount of reversible shrinkage	Surface shortwave absorptivity, thermal conductivity and heat capacity
Chemically stabilized materials (lean concrete, cement treated, soil cement, lime cement flyash, lime flyash, lime stabilized)	Elastic modulus, resilient modulus, Poisson's ratio, unit weight	Minimum resilient modulus, modulus of rupture, base erodibility	Thermal conductivity and heat capacity
Unbound base, subbase and subgrade	Seasonally adjusted resilient modulus, Poisson's ratio, unit weight, coefficient of lateral pressure	Gradation parameters and base erodibility	Plasticity index, gradation parameters, effective grain size, specific gravity, saturated hydraulic conductivity, optimum moisture content, parameters to define soil water characteristic curve (SWCC)

Source: Adapted from: Guide for Mechanistic-Empirical Design of New and Rehabilitated Pavement Structures, Final Report, Part 2. Design Inputs, Chapter 2. Material Characterization, NCHRP, submitted by ARA, Inc., March, 2004)

pavements are susceptible to failure under poor drainage conditions, since the proper function of the PCC layer is dependent on the proper support from the underlying soil and aggregate layers. If these layers weaken as a result of the presence of water, then the PCC slabs undergo excessive deflection, and subsequently fail. Instances of "floating" away of PCC slabs have also been noted because of washing away of supporting layers by severe flooding (Figure 1.3).

Temperature also plays a crucial role in dictating the behavior of the bound layers directly, and the unbound layers indirectly. Generally, HMA layers experience a reduction in stiffness with an increase in the temperature (and vice versa) because of the temperature-dependent rheological characteristics of the asphalt binder. A high stiffness can lead to cracking at low temperatures, whereas a low stiffness can lead to permanent deformation (rutting) at high temperatures. This difference in temperature is generally simulated during laboratory testing for material properties, and the temperature versus stiffness data are considered during the design of a flexible pavement for a specific project location.

PCC layers undergo thermal stresses due to temperature gradient along the depth of the slabs, which can change significantly over the course of a day, depending on the range of temperature at the location of the pavement. The thermal stresses represent a major consideration in the structural design of PCC pavements.

Low temperature in subgrade soils can lead to freezing of water, if any, within the soil. Water in frozen state in the soil layers can lead to heaving in pavements, and subsequently weaken the pavement during thawing. The results from this action are surface roughness and breakage of the surface layer and formation of potholes.

The other major factor affecting the life of the pavement is traffic loading – both load (stress, to be more specific) as well as the number of load repetitions. Generally, the loading and repetitions are estimated on the basis of traffic counts and forecasting and are considered appropriately in the design process. Again it is crucial to make sure that the estimated and predicted parameters are accurate – a deviation of the load from the estimated loading, through say overloading, can lead to premature failure. In this respect, it is important to note that the stress distribution under tires is a very significant factor – major deviation of actual from assumed conditions can lead to premature failures also. For example, high shear stresses can lead to surface cracking, which is generally not considered in conventional ME design.

Apart from the primary causes that influence the life of the pavements, there are a few secondary causes as well. For example, the presence of surface cracks or low-density joint can lead to the ingress of water inside the pavement, which can then weaken the soil and aggregate layers, and subsequently lead to premature deterioration of the entire pavements structure. Furthermore, heavy and repetitive loading, along with the presence of water inside the pavement layers, can lead to a rapid deterioration of the pavement. Then again, if this happens at a high temperature, moisture vapor can also be a major factor in stripping or removal of asphalt binder and fines from HMA and subsequent deterioration of the entire layer. Finally, even if water enters the pavement, it may function well and not suffer, if the water is drained away quickly and if the material is adequately resistant to the impacts of water for the time period during which it is subjected to it and/or the structure is adequately thick to remain unaffected by localized deterioration.

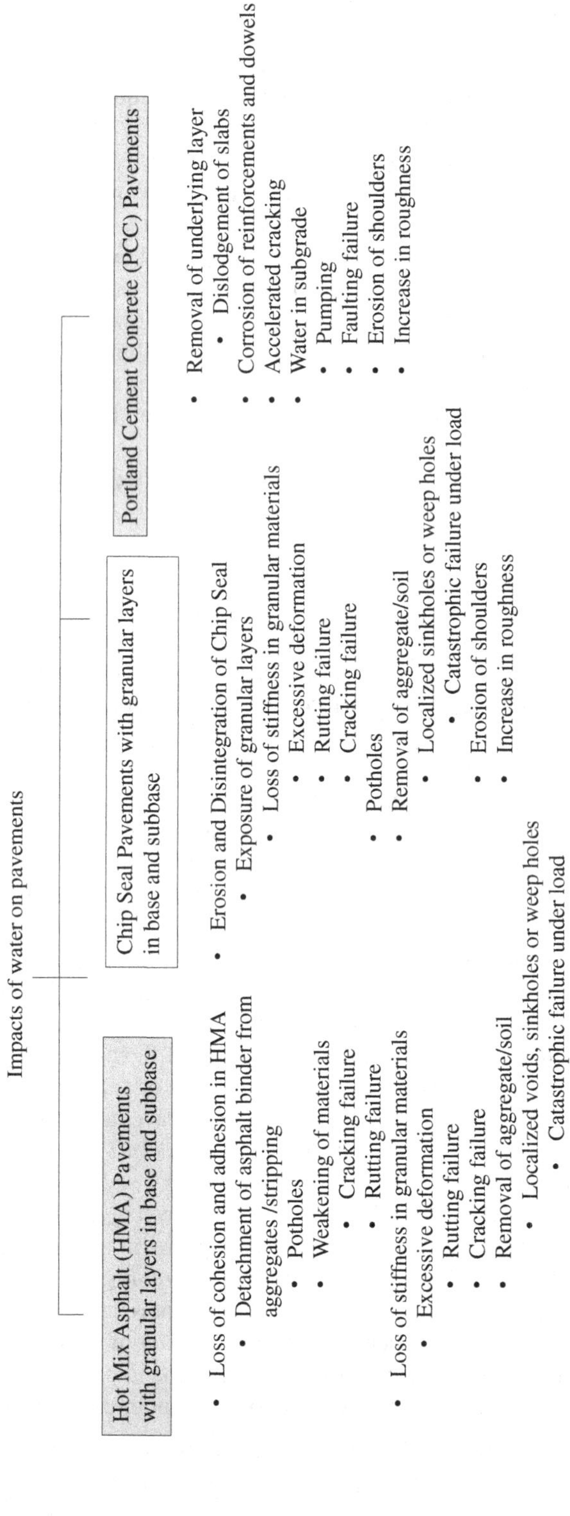

FIGURE 1.3 Impact of water on different types of pavement layers.

Therefore, it can be seen that there exist major interactions between loading, drainage, structure, materials and temperature, that affect the life of pavements significantly. While the variations in temperature can be measured, and hence modeled and predicted with relatively high accuracy, it is the presence of water, its amount and its form (liquid, solid or vapor) within the spatial extent of the pavements structure and the temporal extent of its life, that are difficult to measure and therefore to predict. This is because these factors are not only dependent on the source of water and the intensity and duration of precipitation, but also on the drainage characteristics of layers and their combinations, which are themselves functions of the material properties, and the amounts of water (or saturation) present. Therefore, one of the basic principles of pavement design is to provide for adequate drainage such that the water can be drained away sufficiently quickly under the "worst" condition. It is in defining the "worst" condition that most of the research questions linger. Note that the worst condition may not be necessarily only due to the worst rain – it can be due to a combination of heavy rainfall and low drainage capacity of the pavement plus low moisture resistance of the different layers.

1.5 MIX AND STRUCTURAL DESIGN OF PAVEMENTS

There are primarily two types of design that go hand-in-hand for pavements. One is the mix design and the other is the structural design. In the mix design the optimum constituent materials in a mix (HMA, PCC or recycled base course mix) and their contents are determined on the basis of a set of criteria that ensures the desirable performance of the mix. In structural design, the thicknesses of the different layers are determined on the basis of a set of criteria that ensures the proper performance of each layer, and that of the entire pavement structure, over its design life. Note that the structural design uses mix properties as well as material properties which are obtained from testing of the optimum mix and materials that have already been selected in the mix design step. However, mix design and then structural design is not a one-way method – quite frequently the mix design process needs to be iterated when the structural design, on the basis of the first pass of mix design, is found be impractical or inadequate.

Starting from the ground up, the compaction level of the soil and aggregate layers need to be determined on the basis of optimum moisture content and maximum dry density levels. In general, moisture content versus dry density data are developed on the basis of a Proctor test, and the moisture content corresponding to the maximum dry density is selected as the optimum moisture content. This moisture content and a percentage (such as 95%) of the maximum dry density are specified for construction. Now, for ME structural design, the resilient modulus (M_r) of the soil and aggregate layers are required. The M_r test is conducted by subjecting a cylindrical sample to repetitive sets of loads which produce different sets of stress levels, along with the application of confining stress. The end result of this exercise, which is used in structural design, is the M_r versus stress levels (various forms, such as deviator stress or octahedral stress or bulk stress, depending on the type of material) model, which could include the effect of saturation or moisture content also. A significant amount of research has been conducted on developing accurate models, specifically with respect to the impact of moisture content of saturation – which is directly related to the drainage condition of the pavement. Note that in empirical pavement design the California Bearing Ratio (CBR) (the test is conducted at the fully saturated condition of the material) can be utilized, and there are some equations that relate CBR to M_r as well.

Mix design of HMA (Figure 1.4) is a step-by-step process that considers relevant strength and durability properties of aggregates, performance related properties of the asphalt binder and finally volumetric and durability related properties of the mix. In the widely used Superpave system, for example, abrasion resistance, soundness and gradation (to name a few) are checked for aggregates for specific layers, the Performance Grade (PG) grade of the asphalt binder is ensured for the asphalt for the specific project location, layer and the traffic, and the air voids, voids in mineral aggregate and voids filled with asphalt, as well as retained tensile strength after moisture conditioning are checked for the mix. One crucial step in the process of HMA mix design is the compaction of the samples in the laboratory for the determination of the optimum aggregate gradation and the asphalt content. This compaction process must be as much simulative of field compaction as it is possible, and in the Superpave system, the Superpave Gyratory Compactor (SGC) is used. The number of gyrations in the SGC is related to the level of design traffic (in terms of Equivalent Single Axle Load, ESAL) on the pavement. Note that prior to the introduction of Superpave, the Marshall hammer was used for compaction, and in that case the number of blows was related to the level of design traffic on the pavement.

The structural design process of HMA (Figure 1.4) has evolved significantly over the past decades, from a purely empirical approach to a more sophisticated ME approach. In the empirical approach the design process is based on experience and experimental results, such as the AASHTO '93 structural design method (AASSHTO, 1993) primarily based on the AASHO (American Association of State Highway Officials) road test or the Federal Aviation Administration (FAA, www.faa.gov) empirical method based on tests conducted by the FAA. In this process, the total "cover" that is required over an existing soil subgrade, which is characterized by the CBR value, is read off an equation or a nomograph, and then the cover is split into unbound and bound layers on the basis of minimum thicknesses that are specified for the different layers. Agencies in the different parts of the world have developed similar empirical methods on the basis of local experiments and have used them with varying degrees of success. While these structural design methods are relatively simple, one big disadvantage is that they are not applicable for loading, subgrade/materials and climate conditions that are significantly different from those that were present during the experiments from which the equations or charts have been developed. And over the years, loading and materials have changed considerably all over the world. Another major drawback of the empirical process is the failure in considering the temporal and cyclic changes in environmental conditions in an explicit manner – these changes cause significant variations in some of the key properties of the constituent materials and hence layers in a pavement structure.

The major trend all over the world is now towards ME design. ME design involves the determination of key pavement layer responses (stresses, strains) by using mechanistic equations, and then the estimation of the "life" of the pavement on the basis

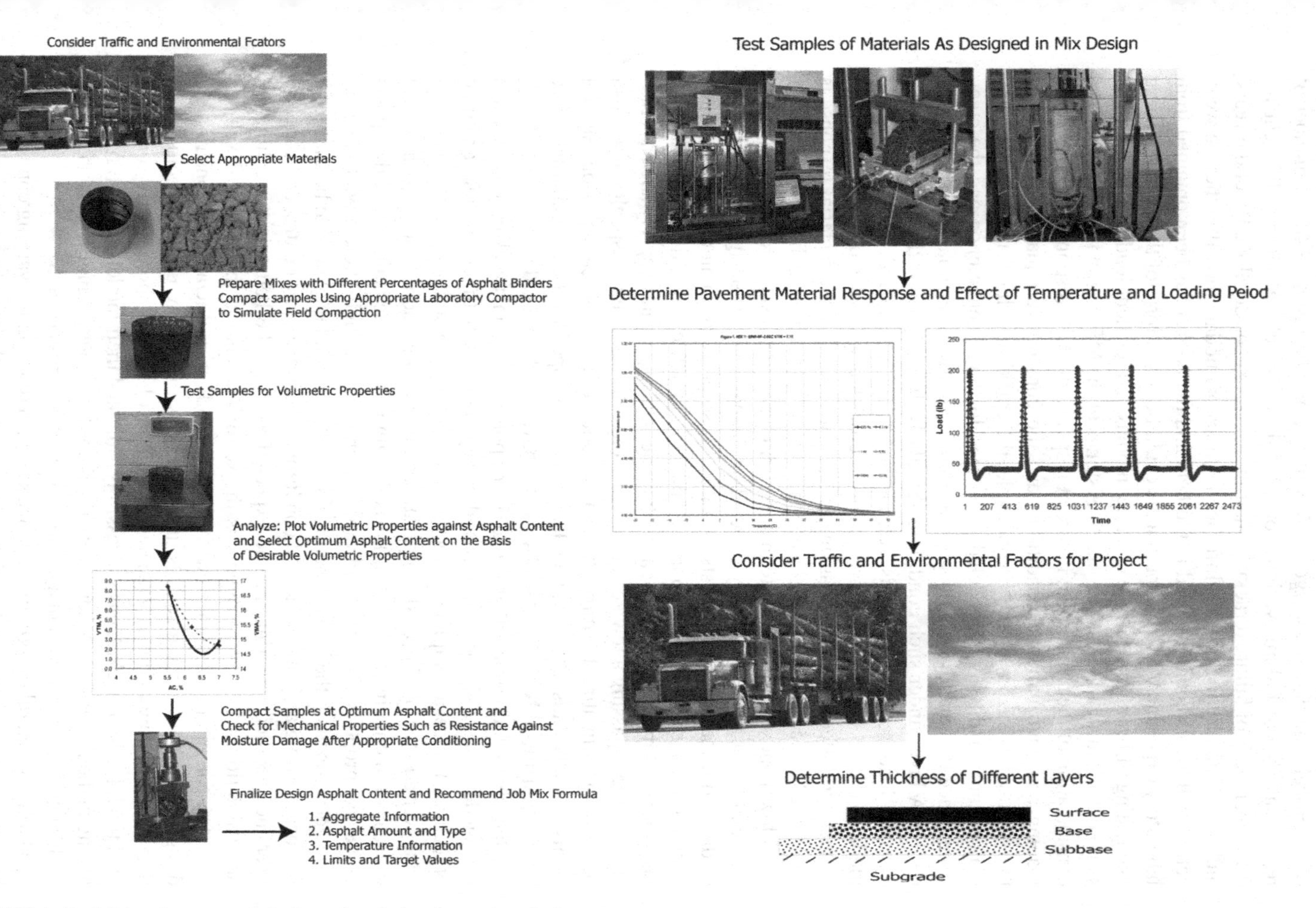

FIGURE 1.4 Mix and structural design of asphalt mix, and asphalt pavement.

of empirical cumulative damage models that relate the response to the number of repetitions (N_f) to failure. Various forms of these models, which are different for different modes of failure (cracking and rutting), have been developed over the years, and they generally include a "shift factor" to accommodate the difference between actual N_f that has been observed in the field and the N_f that has been obtained from laboratory experiments. The experiments could be in the form of third point loading and testing of beams for fatigue cracking models and resilient loading of cylindrical samples for rutting models. The main keystones of the ME design framework are the consideration of the appropriate failure or response model – responses to both traffic loading and environmental conditions, and the constitutive equation of materials that are used in the pavement. For example, the cracking failure of HMA under low temperature conditions is modeled as a creep loading phenomenon and the creep compliance (C_c) and the tensile strength values are used to estimate the response of the HMA layer under such conditions. Similarly, for fatigue cracking, the effect of dynamic loading on a linear viscoelastic material is considered, and the dynamic modulus (E*) is one of the primary mix properties that are utilized. Finally, the high temperature effect is modelled as one of creep loading, and the material response is considered to be that of a viscoelastic material, and properties such as Flow Number and Flow Time are utilized. Note that the E*, C_c and ITS tests need to be conducted on the mix that has been already selected on the basis of mix design. In the absence of data from the appropriate tests of a mix, these properties can also be estimated from the properties of its components and its volumetric properties – for example, viscosity or complex modulus of asphalt binder and aggregate property such as gradation (percent passing a few key sieves) air voids and volumetric asphalt content can be used for estimating E* through the application of an empirical equation. Several forms of such statistical equations have been developed over the years, and more recently, Artificial Neural Networks (ANN) have also been used.

One key factor to remember is that the responses of the material, especially when it is modelled as a viscoelastic material, is dependent on both temperature and time of loading. Hence, a complete characterization usually means the procurement of test data over a range of expected temperatures and time of loading (or frequency, as it is used commonly in HMA testing). Another important issue is the use of the appropriate and, properly calibrated and validated empirical model. These models are generally developed on the basis of national level studies, and they need to be calibrated for local loading and environmental conditions. The greater the control and certainty over the specified material and the expected loading, the better is the calibration and validation. One good example is the set of models that have been developed by FAA for airport design (www.faa.gov/airports_airtraffic/airports/construction/design_software/). FAA materials and layers have relatively unified specifications than those used by highway agencies, and aircraft loads are more tightly controlled than truck loads. The FAA has successfully validated their models through full scale testing in their facility in Atlantic City (William J. Hughes Technical Center), as well as through ongoing studies in a number of airports across the US.

During structural design, the responses to each of the many varying conditions of load and environment are estimated through the use of the appropriate material property, and then the "damage" from all of the conditions are summed up (application of

Miner's rule, Miner, 1945) to estimate the cumulative damage for each year – and the process is repeated for each year for the entire design life (say, 15 years). A design is said to "pass" when the cumulative damage over the design life does not exceed 1 but is close to it. A relatively small number <<1 indicates overdesign whereas a number >1 indicates premature failure and hence under design.

Mix design of concrete (Figure 1.5) materials is general conducted by the absolute volume method (ACI 211, American Concrete Institute, ACI Committee 211, 1991) by proportioning the local concrete materials to achieve the desired fresh and ultimate hardened concrete properties. This includes selecting the water–cement ratio, coarse and fine aggregate, cement and supplementary cementitious materials, water and admixture requirements. The effects of all of these volumetric properties are interrelated. Their selection depends on many factors, including the compressive strength, which is the simplest and easiest to measure. Concrete strength is defined by the average compressive strength of two concrete cylinders. The water-to-cementitious materials ratio (or water–cementitious ratio; W/CM), is the mass of water divided by the mass of all cements, blended cements and pozzolanic materials such as fly ash, slag, silica fume and natural pozzolans. Assuming that a concrete is made with clean sound aggregates, and that the cement hydration has progressed normally, the strength gain is inversely proportional to the water–cementitious ratio by mass. The paste strength is proportional to the solids volume or the cement density per unit volume. The selection of the W/CM is influenced by strength requirements and exposure conditions. The W/CM will govern the permeability or watertightness that is necessary for preventing aggressive salts and chemicals from entering the paste pore system and causing deleterious effects in the concrete or reinforcing steel. Fresh concrete must have the appropriate workability, consistency and plasticity suitable for construction conditions. Workability is a measure of the ease of placement, consolidation and finishing of the concrete. Consistency is the ability of freshly mixed concrete to flow, and plasticity assesses the concrete's ease of molding. If the concrete is too dry and crumbly, or too wet and soupy, then it lacks plasticity. The slump test is a measure of consistency and indicates when the characteristics of the fresh mix have changed or altered. Structural design of concrete (Figure 1.5) can be conducted in different ways. For example, in the PCA method, the design is conducted with respect to two potential failure modes: fatigue and erosion. For fatigue failure, the concrete will fail due to cumulative damage when the stress to strength ratio exceeds a threshold. For erosion, the pavement fails by pumping and erosion of the foundation and joint faulting. The procedure was developed using the results of the finite element computer program JSLAB (Tayabji and Colley, 1983) to determine the critical stresses and deflections induced in concrete pavements by joint, edge and corner loading. The PCA method considers the degree of load transfer provided by dowels or aggregate interlock and the degree of edge support provided by a concrete shoulder. The procedure uses the "composite-k" concept in which the design k is a function of the subgrade soil k, base thickness and base type (granular or cement treated), similar to the AASHTO 1986 and 1993 design procedure.

Fatigue analysis in the PCA method is based on the edge stress midway between transverse joints. The fatigue analysis assumes that approximately 6% of all truck loads will pass sufficiently close to the slab edge to produce a significant tensile

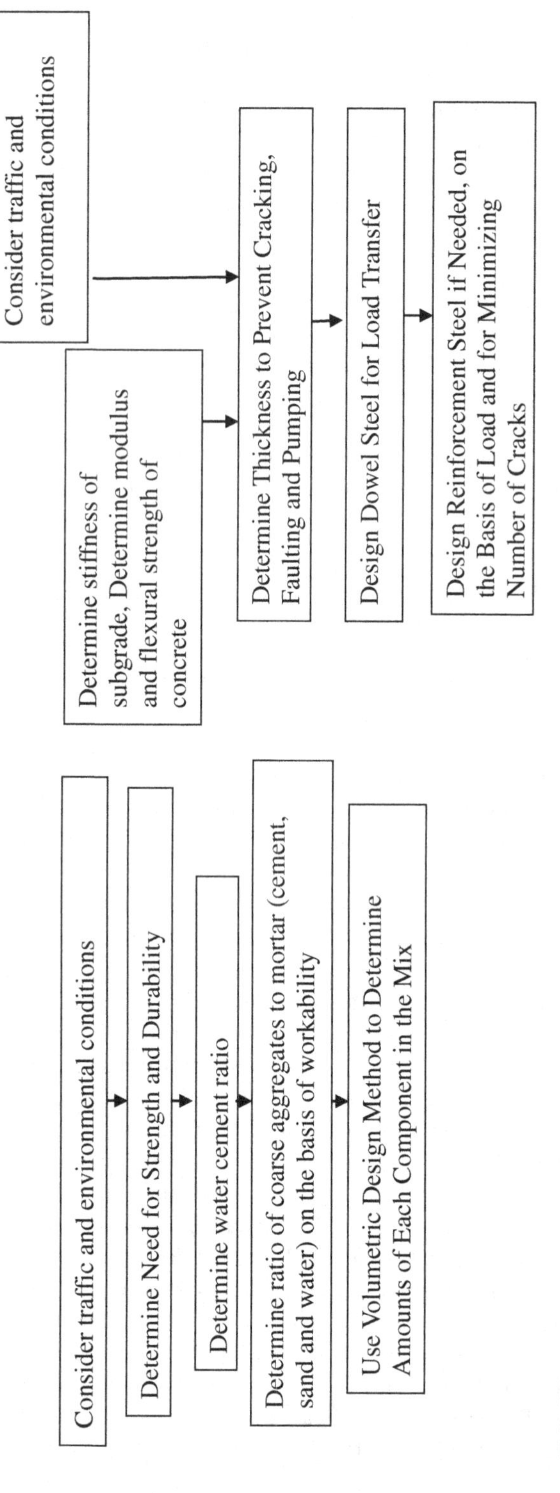

FIGURE 1.5 Mix design of Portland Cement Concrete mix and structural design of concrete pavement.

stress. However, this critical tensile stress is reduced considerably when the main traffic lane and the concrete shoulder are tied. Stresses due to warping and curling are not considered in the PCA method.

1.6 OVERVIEW OF THE ROLE OF WATER IN PAVEMENT PERFORMANCE

It is repeated throughout the literature that the three things most important for satisfactory performance of road pavements are drainage, drainage and drainage. The role of water, in different forms, is very profound in terms of pavement performance (Cedergren, 1974). Water affects almost all materials and layers in a pavement, as discussed below.

In the soil subgrade, particularly for fine grained soils, the modulus values are affected significantly by saturation levels. The permeability of the layer is also affected by the saturation level through matric suction – higher the suction (at lower saturation levels) lower is the permeability. This affects the drainage capability of the layer. Hence, if a project is in a location where there is a significant amount of rainfall, the soil modulus is sensitive to saturation levels, and the drainage of water from the soil subgrade is slow, one can expect a significant deterioration of the structural strength of the pavement due to water.

Subbase and base courses are generally made up of coarse grained materials that are relatively resistant to the impacts of water. However, the presence of large amounts of water can reduce the modulus of these materials significantly, particularly at very high saturation levels. Furthermore, such water can also deteriorate the load bearing capacity of asphalt stabilized base courses, which tend to have relatively high voids and low cohesive strength. Strength of cement stabilized base courses is also reduced over time due to the presence of water.

For both subgrade and subbase/base, the presence of frozen water (ice lenses) can lead to pavement heaves, and their subsequent melting leaves parts of the pavement unsupported – which leads to cracking and formation of potholes under traffic loading.

HMA is affected by water through a process that is commonly referred to as stripping or removal of asphalt binder and fine aggregates from the mix. The mix gradually loses its cohesive strength and hence its ability to withstand tensile strain as well as compressive strain and fails by cracking and/or rutting. The stripping mechanism can be exacerbated by a combined presence of high temperature, water and traffic loading. Load induced high levels of pore pressure in the trapped water underneath the pavement surface can lead to excessive stripping and premature failure of HMA pavements.

The main mechanism by which PCC slabs are affected by water is through ingress of water through the surface or subsurface areas, the formation of slurry of water and underlying layer materials and its subsequent pumping or ejection through joint (mud-pumping). The removal of the underlying material reduces the support of the PCC slab, and as a result the slabs deflect excessively under loading – such excessive deflection leads to premature cracking of the slabs. However, ingress of water into

PCC slabs can also lead to deterioration of steel bars inside the slabs. This is specifically a problem in the case of PCC bridge decks.

Anastasopoulos and Mannering (2015), made the following observation based on a review of 10,000 1-mile long road pavement sections in Indiana:

> An interesting finding is that the condition of pavement drainage (as visually inspected and ranked by the INDOT) plays an important role in pavement deterioration. Inadequate drainage results in faster deterioration of the pavement condition. Therefore, this paper finds that well-drained pavements have higher PCR and lower IRI, RUT, and surface deflection measurements, whereas poorly drained pavements have lower PCR and higher IRI, RUT, and surface deflection measurements.

One more serious effect of water and surface drainage is the reduction in safety and traffic flow capacity of roadways during rainstorms. If the water is not removed quickly, then the presence of water on the pavements surface leads to splash and spray as well as glare from traffic lights under rainy conditions. Presence of water between tires and the pavement surface also lead to loss of friction and hydroplaning. This is also a major problem in airports – particularly high speed movement areas such as runways.

For the latest information on the techniques of limiting water induced damage of pavements the reader is referred to the US National Cooperative Highway Research Program (NCHRP) study: 01-54, Guidelines for Limiting Damage to Flexible and Composite Pavements Due to the Presence of Water, at http://apps.trb.org/cmsfeed/TRBNetProjectDisplay.asp?ProjectID=3626.

REFERENCES

ACI Committee 211. 1991. *Standard Practice for Selecting Proportions for Normal, Heavyweight and Mass Concrete, ACI 211*, pp. 1–91. Farmington Hills, MI: American Concrete Institute.

American Association of State Highway and Transportation Officials (AASHTO). N.d. www.transportation.org.

American Association of State Highway and Transportation Officials (AASHTO). 1986. *Guide for Design of Pavement Structures*, Vol. 2. Washington, DC: AASHTO.

American Association of State Highway and Transportation Officials (AASHTO). 1993. *AASHTO Guide for Design of Pavement Structures*. Washington, DC: AASHTO.

American Concrete Pavement Association (ACPA). 1999. *Survey of States' Concrete Pavement Design and Construction Practices*. Skokie, IL: ACPA.

Anastasopoulos, Panagiotis Ch. and Mannering, Fred L. 2015. Analysis of pavement overlay and replacement performance using random parameters hazard-based duration models. *Journal of Infrastructure Systems*, 21(1): 04014024.

Cedergren, H.R. 1974. *Drainage of Highway and Airfield Pavements*. New York: John Wiley and Sons.

Federal Aviation Administration (FAA). N.d. www.faa.gov.

Federal Aviation Administration (FAA). N.d. Design software. www.faa.gov/airports_airtraffic/airports/construction/design_software/.

Miner, M.A. 1945. Cumulative damage in fatigue. *ASME Transactions*, 67: A159–A164.

Newcomb, D.E. et al. 2001. Concepts of perpetual pavements in perpetual bituminous pavements, Transportation Research Circular, No. 503, Transportation Research Board, National Research Council, Washington, DC.

Smith, P. 1963. Past Performance of Composite Pavements. In *Highway Research Record 37,* Washington, DC: HRB, National Research Council, pp. 14–30.

Tayabji, S.D. and B.E. Colley. 1983. *Improved Pavement Joints, Transportation Research Record 930.* Washington, DC: Transportation Research Board, National Research Council, pp. 69–78.

2 Estimation of Surface Runoff from Storm Water

This chapter provides an overview of the methods that could be utilized for estimating the amount of water that needs to be drained out of a pavement structure. This amount of water is governed by the characteristics of the rainfall and the surface over which the water flows (runoff) before it reaches the drainage structures.

2.1 RATIONAL METHOD

There are various methods of estimating the surface runoff from storm water, the simplest of which is the Rational method, which can be used for areas not exceeding 80 ha. In this method the flow is estimated with the following equation:

$$Q = 0.00278\,CIA \tag{2.1}$$

Where,

Q = the discharge, m^3/s
C = the runoff coefficient
I = the rainfall intensity, mm/h
A = the drainage area, hectares (ha) [1 ha = 10,000 m^2]

The Runoff Coefficient, C, is the ratio of runoff to rainfall. It is affected by the soil group, land use and average land slope. A weighted C value is used if the drainage area consists of different types of surfaces, as follows:

$$Weighted\,C, C_w = \frac{A_1C_1 + A_2C_2 + A_3C_3 \cdots A_nC_n}{A_1 + A_2 + A_3 + \ldots A_n} \tag{2.2}$$

Where,

A_1, A_2, and A_3 = the subareas with different types of surfaces
C_1, C_2, and C_3 = the corresponding runoff coefficients

Typical values of runoff coefficients are shown in Table 2.1.

TABLE 2.1
Typical Runoff Coefficients

Recommended Coefficient of Runoff for Pervious Surfaces by Selected Hydrologic Soil Groupings and Slope Ranges

Slope	A	B	C	D
Flat (0–1%)	0.04–0.09	0.07–0.12	0.11–0.16	0.15–0.20
Average (2–6%)	0.09–0.14	0.12–0.17	0.16–0.21	0.20–0.25
Steep (over 6%)	0.13–0.18	0.18–0.24	0.23–0.31	0.28–0.38

Note: A - Low Runoff potential; B - Moderately Low; C - Moderately High; D - High.
Source: Storm Drainage Design Manual, Erie and Niagara Counties Regional Planning Board.

Recommended Coefficient of Runoff Values for Various Selected Land Uses

Description of Area	Runoff Coefficients
Business: Downtown areas	0.70–0.95
Neighborhood areas	0.50–0.70
Residential: Single-family areas	0.30–0.50
Multi units, detached	0.40–0.60
Multi units, attached	0.60–0.75
Suburban	0.25–0.40
Residential (0.5 ha lots or more)	0.30–0.45
Apartment dwelling areas	0.50–0.70
Industrial: Light areas	0.50–0.80
Heavy areas	0.60–0.90
Parks, cemeteries	0.10–0.25
Playgrounds	0.20–0.40
Railroad yard areas	0.20–0.40
Unimproved areas	0.10–0.30

Source: Hydrology, Federal Highway Administration, HEC No. 19, 1984.

Coefficients for Composite Runoff Analysis

Surface	Runoff Coefficients
Street: Asphalt	0.70–0.95
Concrete	0.80–0.95
Drives and walks	0.75–0.85
Roofs	0.75–0.95

Source: Hydrology, Federal Highway Administraton, HEC No 19, 1984.
Source: From American Association of State Highway and Transportation Officials, *MDM-SI-2, Model Drainage Manual*, 2000 Metric Edition, AASHTO, Washington, DC, © 2000. Used with permission.

The rainfall intensity (I) is obtained from Intensity-Duration-Frequency (IDF) curves for specific locations. The time of concentration (T_c) and the return period need to be determined or estimated in order to utilize the IDF curves and is used in the determination of inlet spacing as well as pipe sizing. T_c is the time required for the water to flow from the hydraulically most distant point of the drainage area to the inlet (inlet time). T_c is calculated as the sum of the time required for water to flow across the pavement of overland back of the curb to the gutter plus the time required for flow to move through the length of the gutter to the inlet. A minimum T_c of 7 minutes should be used if the total time of concentration to the upstream inlet is less than 7 minutes. For pipe sizing, the T_c consists of the inlet time plus the time required for the water to flow through the storm drain to the point under consideration. If there is more than one source of runoff to a given point in a drain system, the longest T_c is used. A minimum T_c of 5 minutes is recommended for municipal areas.

The time to flow overland is estimated by using distance-slope-time to travel relations as shown in Figure 2.1, or by using the velocity-versus slope relation for different types of surfaces as shown in Figure 2.2. The time required to flow within a storm drain (channel) is estimated by using Manning's equation, given in Chapter 4.

The design frequency and return period (RP) or recurrence interval (RI) are used to estimate the likelihood or probability of a storm with a specific Intensity and duration of rainfall. The frequency with which a given flood can be expected to occur is the reciprocal of the probability that the flood will be equaled or exceeded in a given year. For example, if a flood has a probability of 5% of being equaled or exceeded each year, over a long period of time, then the frequency is 1/(5/100) = 20, which means that the flood will be equaled or exceeded on an average of once in every 20 years. The RI or RP is also 20 years. The probability of exceeding this storm is 100/RP. For design purposes, the recommended RP values are shown in Table 2.2.

Rainfall intensity, I, is the intensity of rainfall in mm/h for a duration equal to the time of concentration, and when multiplied by the duration of the rainfall gives the total amount of rain. The IDF curve can be generated from climatic data and weather databases available from various sources such as the NOAA Hydrometeorological Design Studies Center, Precipitation Frequency Data Server (www.nws.noaa.gov/oh/hdsc/).

The Precipitation Frequency Data Server (PFDS) based on NOAA Atlas 14 provides a web portal for estimating precipitation and frequency. The process consists of the following steps (see also Figure 2.3):

1. Access the PFDS homepage (https://hdsc.nws.noaa.gov/hdsc/pfds/).
2. Click on the specific state (for the project location) on the US map.
3. Select the specific project location (latitude, longitude or city street, for example) from the state map.
4. Click on the "Precipitation frequency" tag on the side menu to download the information.

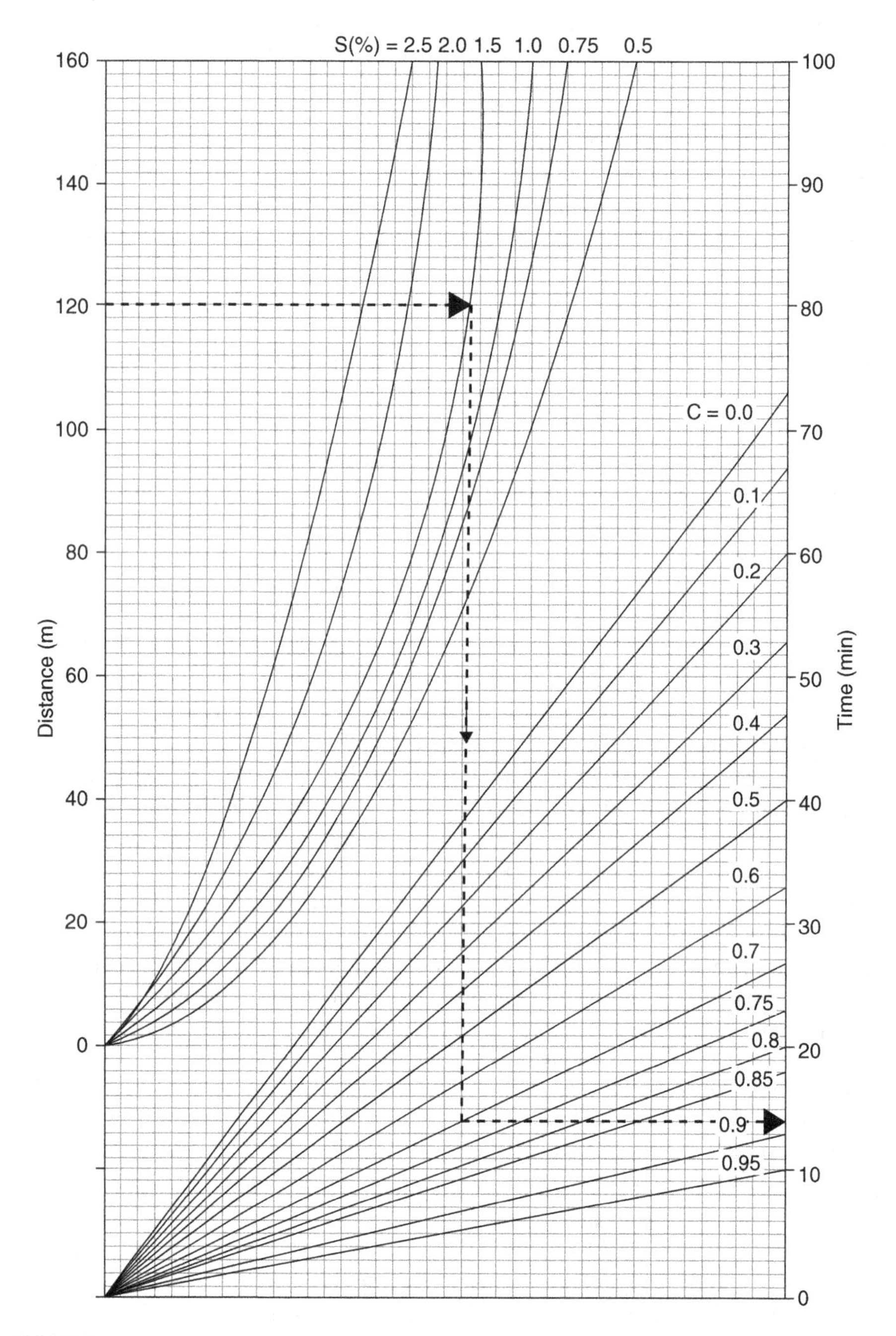

FIGURE 2.1 Direct estimation of overland time of flow. (From American Association of State Highway and Transportation Officials, *MDM-SI-2, Model Drainage Manual*, 2000 Metric Edition, AASHTO, Washington, DC, © 2000. Used with permission.)

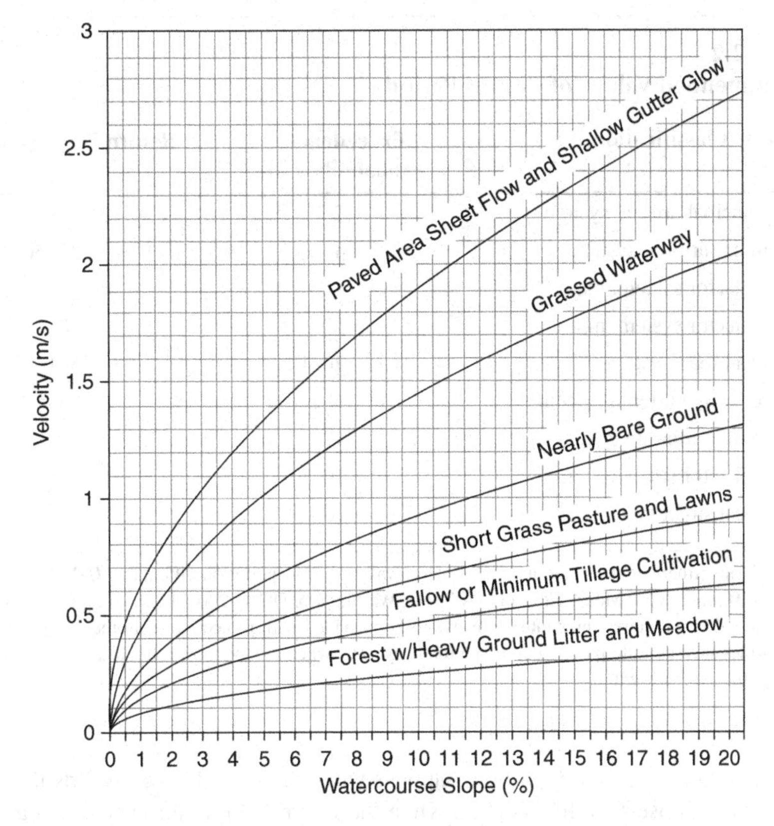

FIGURE 2.2 Estimation of velocity of flow to compute overland time of flow. (From American Association of State Highway and Transportation Officials, *MDM-SI-2, Model Drainage Manual*, 2000 Metric Edition, AASHTO, Washington, DC, © 2000. Used with permission.)

5. The window allows the user to download the information as IDF (Intensity-Duration-Frequency) cures as well as precipitation magnitude frequency curve with upper and lower confidence limits for a specific duration.

2.2 NRCS CURVE NUMBER (CN) METHOD (NRCS, 2004)

A unit hydrograph can be used to estimate peak discharge from a drainage area that is not greater than 2,000 mi². A hydrograph shows discharge versus time. A unit hydrograph shows a plot of time (*x* axis) versus runoff (*y* axis) (Figure 2.4) caused by a 1 inch deep water that is uniformly distributed over the whole drainage area for a given storm and duration of rainfall. Slopes of unit hydrograph caused by storms that are of similar duration (not same intensity) are similar. The ordinates in the unit hydrograph are approximately proportional to the runoff volumes.

TABLE 2.2
Recommended Values of Return Period

Roadway Classification	Exceedence Probability (%)	Return Period (Years)
Rural principal arterial system	2	50
Rural minor arterial system	4–2	25–50
Rural collector system, major	4	25
Rural collector system, minor	10	10
Rural local road system	20–10	5–10
Urban principal arterial system	4–2	25–50
Urban minor arterial street system	4	25
Urban collector street system	10	10
Urban local street system	20–10	5–10

Source: From American Association of State Highway and Transportation Officials, *MDM-SI-2, Model Drainage Manual*, 2000 Metric Edition, AASHTO, Washington, DC, © 2000. Used with permission.
Note: Federal law requires interstate highways to be provided with protection from the 2% flood event and facilities such as underpasses, depressed roadways, etc. where no overflow relief is avilable should be designed for the 2% event.

Surface runoff or overland flow happens when the rainfall rate exceeds the infiltration rate. Subsurface flow occurs when the infiltrated rainfall encounters a layer of lower hydraulic conductivity and travels laterally over the boundary of that layer, reappearing as seep or spring. Channel runoff refers to that which occurs when rain falls on a flowing stream.

On the hydrograph, channel runoff appears at the start of the storm, remains throughout the storm and varies with rainfall intensity. The surface runoff appears on hydrograph once the demands of the interception, infiltration and surface storages are met. It varies during the storm and ends during or soon after the storm. The subsurface flow contributes to the hydrograph during or soon after the storm. The channel, surface and subsurface flow are combined into direct runoff, which is estimated in the NRCS method by the Curve Number (CN) method. The CN is based on soil permeability, surface cover, hydrologic condition and antecedent moisture. The dimensionless unit hydrograph is based on the drainage area time of concentration.

The Curve Number (CN) runoff equation is:

$$Q = \frac{(P - I_a)^2}{(P - I_a) + S} \; for \; P > I_a \tag{2.3}$$

$$Q = 0 \; for \; P \le I_a \tag{2.4}$$

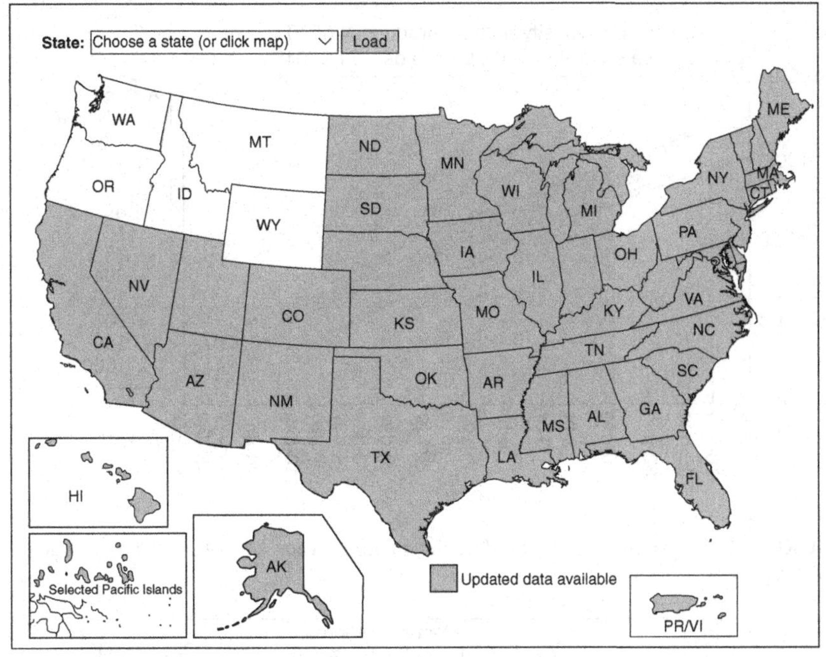

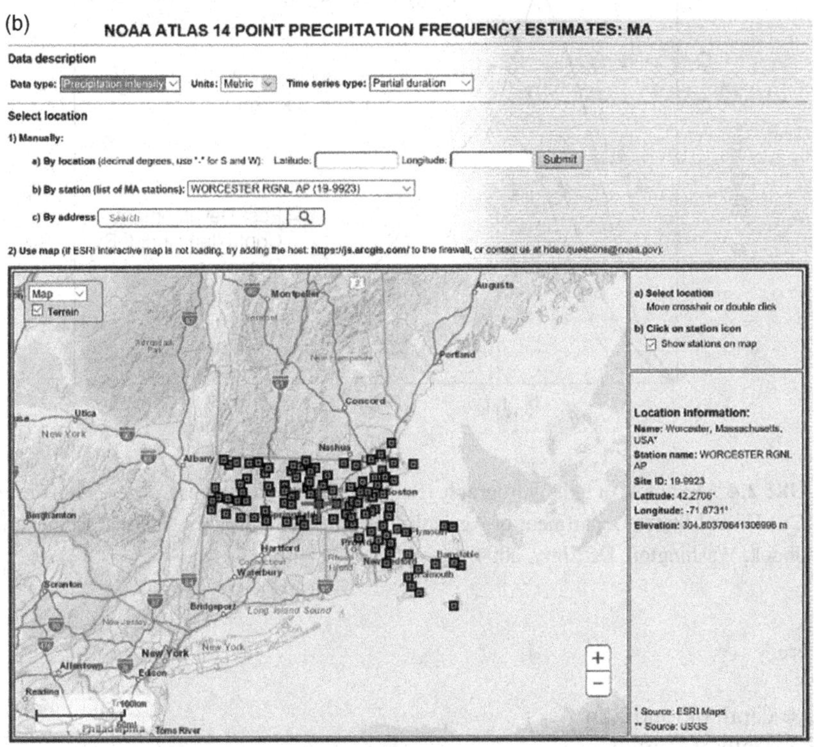

FIGURE 2.3 Steps in generating IDF information from NOAA website: (a) select state from the US map; (b) select the specific station (continued over the page).

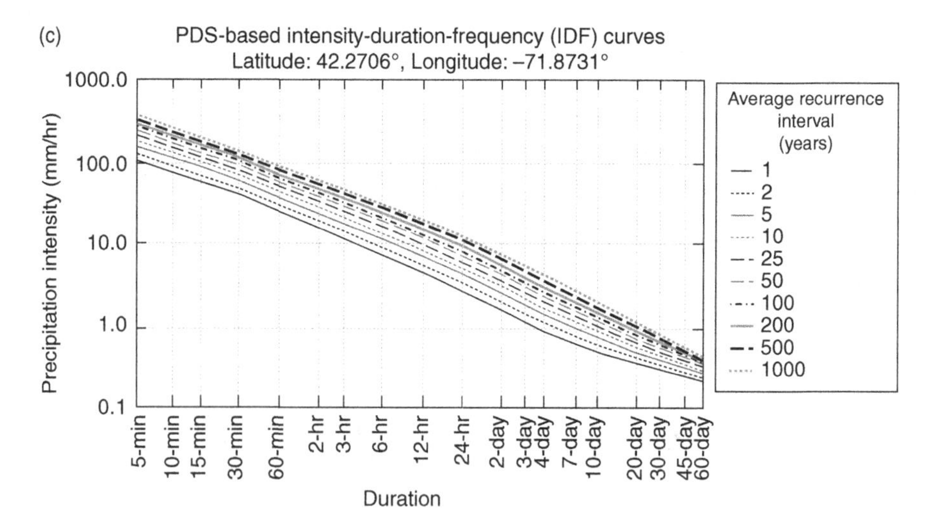

FIGURE 2.3 Steps in generating IDF information from NOAA website: (c) generate IDF curves.

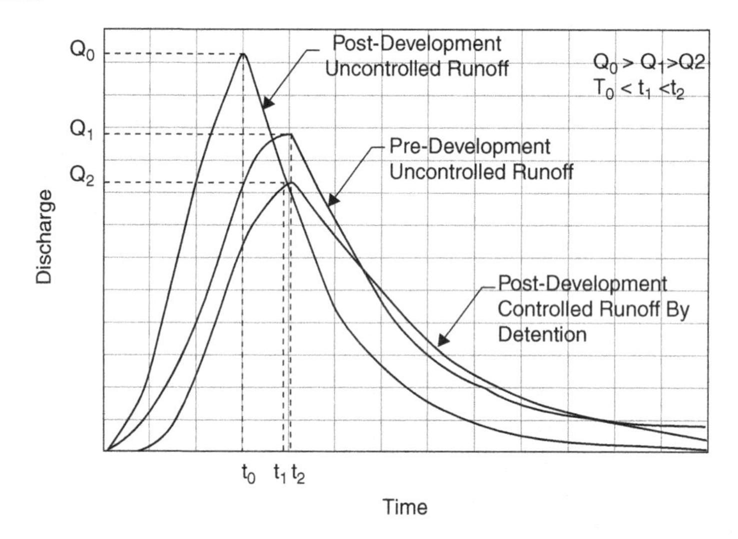

FIGURE 2.4 Example of unit hydrograph. (Source: Natural Resources Conservation Service (NRCS) United States Department of Agriculture Part 630 Hydrology National Engineering Handbook. Washington, DC, July, 2004.)

Where,

Q = depth of runoff, in
P = depth of rainfall, in
I_a = initial abstraction, in
S = maximum potential retention, in

I_a is due to interception, infiltration during early part of the storm and surface depression storage. Infiltration during the early part of the storm depends on rainfall intensity, soil crusting and soil moisture.

Empirical relationship between I_a and S:

$$I_a = 0.2S \tag{2.5}$$

Substituting (2.5) in (2.3) gives

$$Q = \frac{(P-0.2S)^2}{P+0.8S} \tag{2.6}$$

For $P > I_a$ this equation is the basis for Curve Numbers. Equations have been put into a nomograph (Figure 2.5) by calculating Curve Number (CN) as follows:

$$CN = \frac{1000}{10+S} \text{ when } S \text{ is in inch} \tag{2.7}$$

$$CN = \frac{1000}{10+\frac{S}{25.4}} \text{ when } S \text{ is in mm} \tag{2.8}$$

Note:
S is dependent on soil cover and is independent of the storm.
I_a can be considered as the boundary between storm size that produces runoff and the storm size that produces no runoff
Maximum possible loss = I_a + S
F in Figure 2.5 is the actual retention for a storm >I_a
Total actual retention = I_a + F
Note that S does not include I_a

The variability in the CN results are due to rainfall intensity and duration, total rainfall, soil moisture content, cover density, stage of growth and temperature. These causes of variability are collectively called "Antecedent Runoff Condition (ARC)", which is divided into three classes: Class I for dry condition; II for average condition; and III for wet conditions. Efforts to quantitatively explain the scatter have been made on the basis of antecedent soil moisture, which is indicated by 5 day antecedent precipitation; however, no apparent relationship has been found to exist between CN and antecedent precipitation. Another approach is to consider CN as a random number. In this approach the log normal probability distribution of S is computed. Mean of the logarithm corresponds to the ARC II CN. The CN associated with 10% and 90% can be used to express the extreme values of the CN distribution.

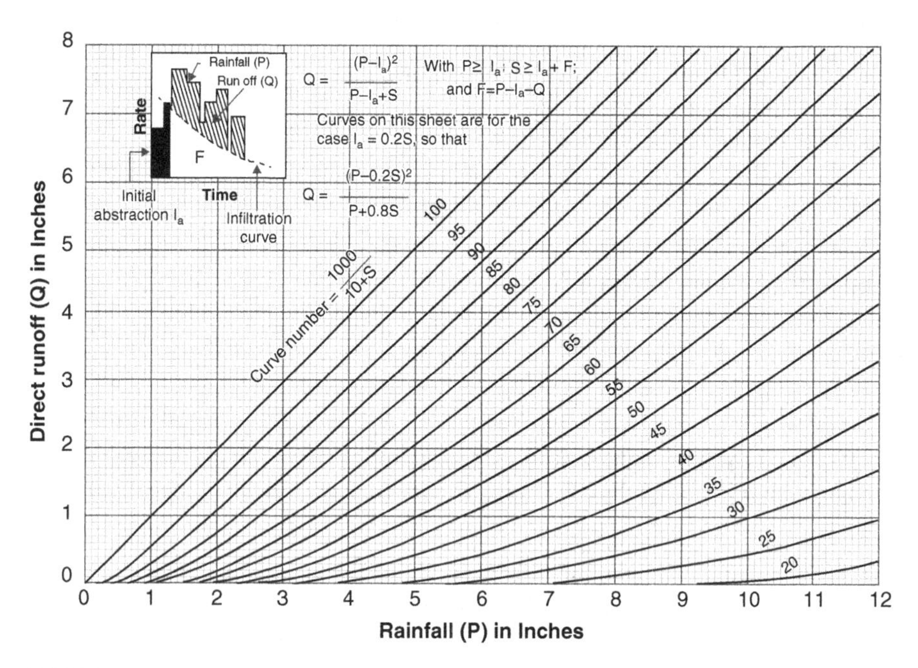

FIGURE 2.5 Graphical solution for NRCS Curve Number method; graphical solution of equation 2.7. (Source: Natural Resources Conservation Service (NRCS) United States Department of Agriculture Part 630 Hydrology National Engineering Handbook. Washington, DC, July, 2004.)

Examples of application

Example 1. Estimate the direct runoff for a storm event with an average depth of 4.3 inch rainfall for a watershed with a pasture in good condition and soil from hydrologic group C

Step 1. Determine CN: See Table 2.3 for pasture, good condition and soil group C, CN = 74; Then see Table 2.4 or CN equation, S = 3.51

Step 2. Estimate runoff; use Figure 2.5; P = 4.3 inch; interpolate for CN = 74, Q = 1.82 inch or use equation 2.7

$$Q = \frac{(P-0.2S)^2}{P+0.8S}$$

Q = 1.82 inch

Example 2. Estimate Q for ARC II and III and compare with that of ARC I

1. Determine CN for ARC II, CN = 74
2. Determine CN for ARC I and III

From Table 2.4, corresponding to CN = 74 for ARC II find CN ARC I = 55, CN ARC III = 88

3. Use Figure 2.5 to get Q

ARC	CN	Direct Runoff	Direct Runoff, Q	
			As % of runoff	As % of Q for ARC II
I	55	0.65	15.1	35.7
II	74	1.82	42.3	100
III	88	3.01	70.0	165

Note that runoff is not proportional to the CN. Direct runoff for watersheds with more than one hydrologic soil cover complex can be computed either by the weighted average of the runoff (Q) or by the weighted average of the CN – the latter is preferred because of relatively simplicity.

Example 3: Use of weighted Q

Watershed area = 630 acres; 400 acre is row crop. Contoured, good condition, 230 acres in rotation meadow, contoured, good rotation; hydrologic soil group B, compute direct runoff. Rainfall P = 5.1 inch, watershed in ARC II.

1. Find CN: Table 2.3 row crop CN = 75, meadow CN = 69
2. Estimate runoff for each complex; use Figure 2.5, rain = 5.1 inch at CN = 75, 69, Q = 2.53 inch, Q = 2.03 inch, respectively.
3. Calculate weighted runoff

Cover	Area (acres)	Q (inch)	Acres * Q
Row crop	400	2.53	1012
Meadow	230	2.03	467
Total	630		1479

$$Q_{WEIGHTED} = \frac{1479}{630} = 2.35\,inch$$

Or, do it by weighted CN

Cover	Area (acres)	CN	Acres * CN
Row crop	400	75	30000
Meadow	230	69	15870
Total	630		45870

$$CN_{WEIGHTED} = \frac{45870}{630} = 72.8,\ use\ 73$$

TABLE 2.3
Runoff Curve Numbers for Agricultural Lands[1]

Cover Description			CN for Hydrologic Soil Group			
Cover Type	**Treatment[2]**	**Hydrologic Condition[3]**	**A**	**B**	**C**	**D**
Fallow	Bare soil	–	77	86	91	94
	Crop residue cover (CR)	Poor	76	85	90	93
		Good	74	83	88	90
Row crops	Straight row (SR)	Poor	72	81	88	91
		Good	67	78	85	89
	SR + CR	Poor	71	80	87	90
		Good	64	75	82	85
	Contoured (C)	Poor	70	79	84	88
		Good	65	75	82	86
	C + CR	Poor	69	78	83	87
		Good	64	74	81	85
	Contoured & terraced (C & T)	Poor	66	74	80	82
		Good	62	71	78	81
	C & T + CR	Poor	65	73	79	81
		Good	61	70	77	80
Small grain	SR	Poor	65	76	84	88
		Good	63	75	83	87
	SR + CR	Poor	64	75	83	86

		Good	60	72	80	84
	C	Poor	63	74	82	85
		Good	61	73	81	84
	C + CR	Poor	62	73	81	84
		Good	60	72	80	83
	C & T	Poor	61	72	79	82
		Good	59	70	78	81
	C & T +CR	Poor	60	71	78	81
		Good	58	69	77	80
Close-seeded or broad castlegumes or rotation meadow	SR	Poor	66	77	85	89
		Good	58	72	81	85
	C	Poor	64	75	83	85
		Good	55	69	78	83
	C & T	Poor	63	73	80	83
		Good	51	67	76	80
Pasture, grassland, or range-continuous forage for grazing[4]		Poor	68	79	86	89
		Fair	49	69	79	84
		Good	39	61	74	80
Meadow-continuous grass, protected from grazing and generally mowed for hay		Good	30	58	71	78

(continued)

TABLE 2.3
(Cont.)

Cover Description			CN for Hydrologic Soil Group			
Cover Type	**Treatment[2]**	**Hydrologic Condition[3]**	**A**	**B**	**C**	**D**
Brush-brush-forbs-grass mixture with brush the major element[5]		Poor	48	67	77	83
		Fair	35	56	70	77
		Good	30[6]	48	65	73
Woods-grass combination (orchard or tree farm)[7]		Poor	57	73	82	86
		Fair	43	65	76	82
		Good	32	58	72	79
Woods[8]		Poor	45	66	77	83
		Fair	36	60	73	79
		Good	30	55	70	77
Farmastead-buildings, lanes, driveways, and surrounding lots		–	59	74	82	86
Roads (including right-of-way)						
Dirt		–	72	82	87	89
Gravel		–	76	85	89	91

Notes:

1 Average runoff condition, and Ia = 0.2s.

2 Crop residue cover applies only if residue is on at least 5% of the surface throughout the year.

3 Hydrologic condition is based combinations of factors that affect infiltration and runoff, including: (a) density and canopy of vegetative areas; (b) amount of year-round cover; (c) amount of grass or close-seeded legumes; (d) percent of residue cover on the land surface (good ≥20%); and (e) degree of surface toughness.

Poor: Factors impair infiltration and tend to increase runoff.

Good: Factors encourage average and better then average infiltration and tend to decrease runoff.

For conservation tillage poor hydrologic condition, 5–20% of the surface is covered with residue (less than 750 pounds per acre for row crops or 300 pounds per acre for small grain).

For conservation tillage good hydrologic condition, more than 20% of the surface is covered with residue (greater than 750 pounds per acre for row crops or 300 pounds per acre for small grain).

4 Poor: <50% ground cover or heavily grazed with no mulch.

Fair: 50–75% ground cover and not heavily grazed.

Good: >75% ground cover and lightly or only occasionally grazed.

5 Poor: <50% ground cover.

Fair: 50 to 75% ground cover.

Good: >75% ground cover.

6 If actual Curve Number (CN) is less than 30, use CN = 30 for runoff computation.

7 CNs shown were computed for areas with 50% woods and 50% grass (pasture) cover. Other combinations of conditions may be computed from the CNs for woods and pasture.

8 Poor: Forest litter, small trees, and brush are destroyed by heavy grazing or regular burning.

Fair: Woods are grazed, but not burned, and some forest litter covers the soil.

Good: Woods are protected from grazing, and litter and brush adequately cover the soil.

Source: Natural Resources Conservation Service (NRCS) United States Department of Agriculture Part 630 Hydrology National Engineering Handbook. Washington, DC, July, 2004.

TABLE 2.4
Curve Numbers and Constants for the Case $I_a = 0.2S$

1	2	3	4	5	1	2	3	4	5
CN for ARC II	CN for ARC		S Values* (in)	Curve* Starts Where P – (in)	CN for ARC II	CN for ARC		S Values* (in)	Curve* Starts Where P – (in)
	I	III				I	III		
100	100	100	0	0	60	40	78	6.67	1.33
99	97	100	0.101	0.02	59	39	77	6.95	1.39
98	94	99	0.204	0.04	58	38	76	7.24	1.45
97	91	99	0.309	0.06	57	37	75	7.54	1.51
96	89	99	0.417	0.08	56	36	75	7.86	1.57
95	87	98	0.256	0.11	55	35	74	8.18	1.64
94	85	98	0.638	0.13	54	34	73	8.52	1.70
93	83	98	0.753	0.15	53	33	72	8.87	1.77
92	81	97	0.870	0.17	52	32	71	9.23	1.85
91	80	97	0.989	0.20	51	31	70	9.61	1.92
90	78	96	1.11	0.22	50	31	70	10.0	2.00
89	76	96	1.24	0.25	49	30	69	10.4	2.08
88	75	95	1.36	0.27	48	29	68	10.8	2.16
87	73	95	1.49	0.30	47	28	67	11.3	2.26
86	72	94	1.63	0.33	46	27	66	11.7	2.34
85	70	94	1.76	0.35	45	26	35	12.2	2.44
84	68	93	1.90	0.38	44	25	64	12.7	2.54
83	67	93	2.05	0.41	43	25	63	13.2	2.64
82	66	92	2.20	0.44	42	24	62	13.8	2.76
81	64	92	2.34	0.47	41	23	61	14.4	2.88
80	63	91	2.50	0.50	40	22	60	15.0	3.00
79	62	91	2.66	0.53	39	21	59	15.6	3.12
78	60	90	2.82	0.56	38	21	58	16.3	3.26
77	59	89	2.99	0.60	37	20	57	17.0	3.40
76	58	89	3.16	0.63	36	19	56	17.8	3.56
75	57	88	3.33	0.67	35	18	55	18.6	3.72
74	55	88	3.51	0.70	34	18	54	19.4	3.88
73	54	87	3.70	0.74	33	17	53	20.3	4.06
72	53	86	3.89	0.78	32	16	52	21.2	4.24
71	52	86	4.08	0.82	31	16	51	22.2	4.44

TABLE 2.4
(Cont.)

1	2	3	4	5	1	2	3	4	5
CN for ARC II	CN for ARC		S Values* (in)	Curve* Starts Where P – (in)	CN for ARC II	CN for ARC		S Values* (in)	Curve* Starts Where P – (in)
	I	III				I	III		
70	51	85	4.28	0.86	30	15	50	23.3	4.66
69	50	84	4.49	0.90	25	12	43	30.0	6.00
68	48	84	4.70	0.94	20	9	37	40.0	8.0
67	47	83	4.92	0.98	15	6	30	56.7	11.34
66	46	82	5.15	1.03	10	4	22	90.0	18.00
65	45	82	5.38	1.08	5	2	13	190.0	38.00
64	44	81	5.62	1.12	0	0	0	Infinity	Infinity
63	43	80	5.87	1.17					
62	42	79	6.13	1.23					
61	41	78	6.39	1.28					

Note: * For CN in column 1.

Source: Natural Resources Conservation Service (NRCS) United States Department of Agriculture Part 630 Hydrology National Engineering Handbook. Washington, DC, July, 2004. Used with permission.

Use Figure 2.5, rain of 5.1 inch, CN = 73, Q = 2.36 inch

Note that the weighted Q gives the correct result although it takes more time than the weighted CN method. The difference between the two methods is the largest in watershed with widely differing CNs and lower rainfalls.

REFERENCES

NOAA Hydrometeorological Design Studies Center, Precipitation Frequency Data Server. N.d. www.nws.noaa.gov/oh/hdsc.

Natural Resources Conservation Service (NRCS). 2004. *United States Department of Agriculture Part 630 Hydrology National Engineering Handbook*. Washington, DC.

3 Estimation of Subsurface Water

In general, there are four sources of subsurface water:

1. Ground water or water table.
2. Artesian aquifer.
3. Water resulting from melting of ice.
4. Water that infiltrates through voids and cracks in the upper layers.

If the groundwater or the water table is not lowered sufficiently before the construction of the pavement, then the total inflow from it to the collector pipe (q_d) and the drainage layer (q_g) can be estimated first by calculating the radius of influence (L_i) and the by estimating the inflow from the bottom of the drainage layer and that from the area above the drainage layer, as illustrated in Figure 3.1. The following equations are used in the calculations:

$$L_i = 3.8(H - H_0) \tag{3.1}$$

$$q_1 = \frac{k(H - H_0)}{2L_1} \tag{3.2}$$

Where, q_1 is the flow from above the drainage layer, and H and H_0 as defined in Figure 3.1.

$$q_d = q_1 + q_2 \tag{3.3a}$$

Where q_d is the total lateral inflow into the drainage pipe for collector pipes on *both sides of the road* and q_2 is the flow coming from the bottom of the drainage layer as determined from Figure 3.1 and k, the permeability of the subgrade material.

$$q_d = 2(q_1 + q_2) \tag{3.3b}$$

Where q_d is the total lateral inflow into the drainage pipe for a collector pipe on *one side of the road*:

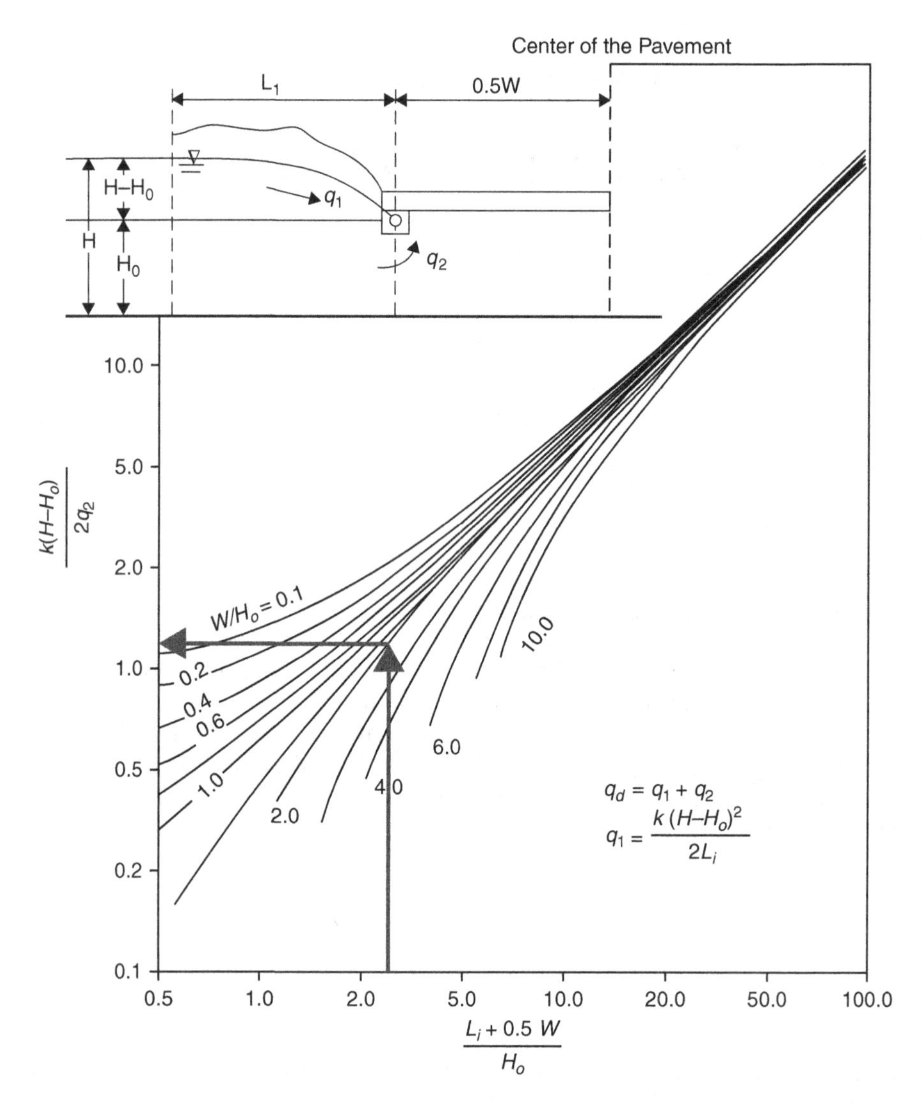

FIGURE 3.1 Estimation of inflow from groundwater (From Garber, N.J. and Hoel, L.A., *Traffic and Highway Engineering*, 3rd edn., Thomson Learning, Clifton Park, NY. © 2002, Nelson Education Ltd., Reproduced by permission. www.cengage.com/permissions.)

$$q_g = \frac{2q_2}{W} \tag{3.4a}$$

Where q_g is the groundwater flow per unit area into drainage layer of a pavement *sloped and with collector pipes on both sides* and W is the width of the pavement:

$$q_g = \frac{(q_1 + 2q_2)}{W} \tag{3.4b}$$

Where q_g is the groundwater flow per unit area into drainage layer of a pavement *sloped and with a collector pipes on one side.*

The following equation is used to estimate the seepage water from artesian aquifers into a subsurface drainage layer:

$$q_a = K\frac{\Delta H}{H_0} \tag{3.5}$$

q_a = the inflow from artesian source, $ft^3/day/ft^2$ of drainage area
ΔH = the excess hydraulic head, ft
H_0 = the thickness of the subgrade soil between the drainage layer and the artesian aquifer, ft
K = the coefficient of permeability, ft/day

To calculate the water from melting snow, the average rate of heave or the frost susceptibility classification of the subgrade soil and the overburden pressure on the soil (from the surface, base and subbase layers) are utilized, as explained in Figure 3.2. Table 3.1 shows the frost susceptibility of different types of soils.

The amount of water Infiltration through joints and cracks is estimated by the following equation.

$$q_i = I_c[\frac{N_c}{W} + \frac{W_c}{WC_s}] + k_p \tag{3.6}$$

Where,

q_i = the rate of pavement infiltration, $m^3/day/m^2$ ($ft^3/day/ft^2$)
I_c = the crack infiltration rate, $m^3/day/m$ ($ft^3/day/ft$); suggested value, (I_c) = 0.223 $m^3/day/m$, 2.4 $ft^3/day/ft$ of crack (Ridgeway, 1976)
N_c = the number of longitudinal cracks
W_c = the length of contributing transverse joints or cracks, m (ft)
W = the width of permeable base, m (ft)
C_s = the spacing of contributing transverse joints or cracks, m (ft)
k_p = the pavement permeability, m/day (ft/day); suggested permeability of uncracked specimens of asphalt pavement (AC) after being subjected to traffic and Portland Cement Concrete pavement (PCC) are on the order of 1×10^{-9} cm/s (15×10^{-5} ft/day) (Barber and Sawyer, 1952)

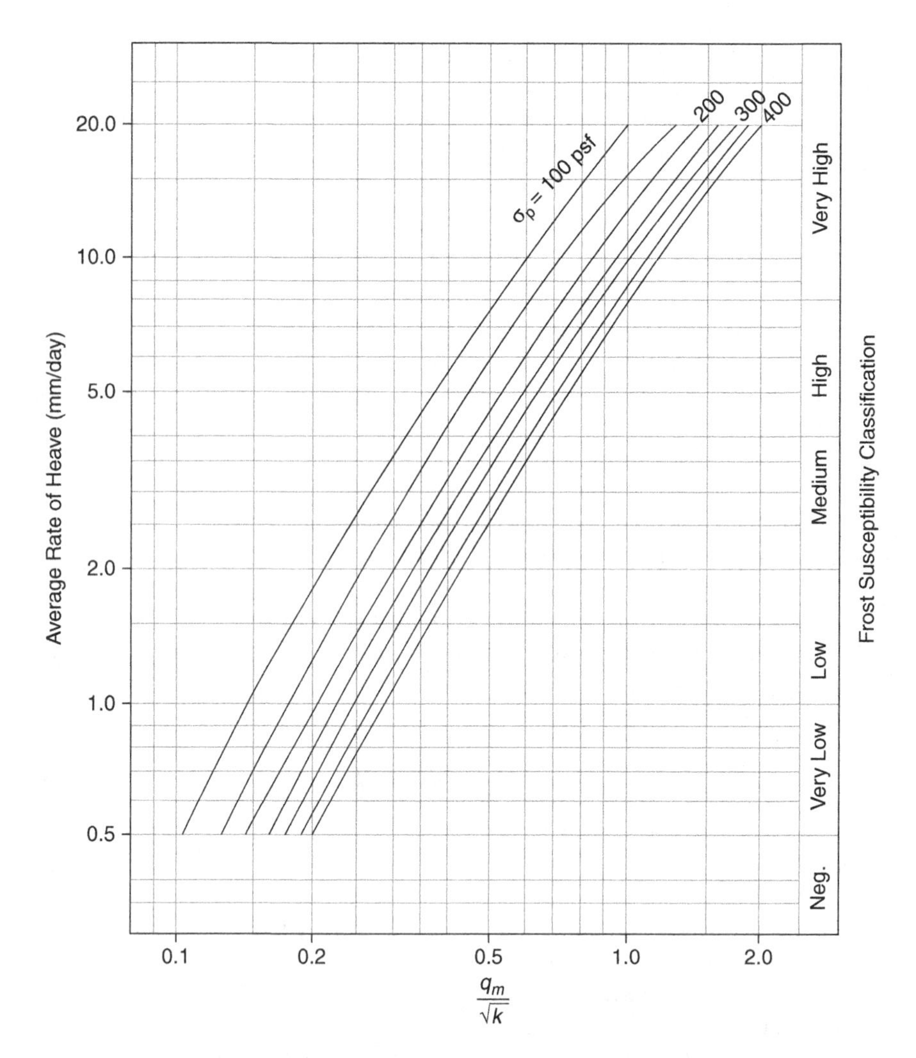

FIGURE 3.2 Chart for estimating inflow due to melting snow. (From Moulton, L.K., Highway subsurface design, FHWA-TS-80–224, U.S. Department of Transportation, Washington, DC, 1980.)

TABLE 3.1
Frost Susceptibility of Different Types of Soils

Unified Classification	Symbol	Percent <0.02 mm	Heave Rate mm/day	Frost Susceptibility
Gravels and sand gravels	GP	0.4	3.0	Medium
Gravels and sand gravels	GW	0.7–1.0	0.3–1.0	Low
Gravels and sand gravels	GW	1.0–1.5	1.0–3.5	Low to medium
Gravels and sand gravels	GW	1.5–4.0	3.5–2.0	Medium
Silty and sandy gravels	GP – GM	2.0–3.0	1.0–3.0	Low to medium
Silty and sandy gravels	GW – GM & GM	3.0–.0	3.0–4.5	Medium to high
Clayey and silty gravels	GW – GC	4.2	2.5	Medium
Clayey and silty gravels	GM – GC	15.0	5.0	High
Clayey and silty gravels	GC	15.0–30.0	2.5–5.0	Medium to high
Sands and gravelly sands	SP	1.0–2.0	0.8	Very low
Silty and gravelly sands	SW	2.0	3.0	Medium
Silty and gravelly sands	SP – SM	1.5–2.0	0.2–1.5	Low
Silty and gravelly sands	SW – SM	2.0–5.0	1.5–6.0	Low to high
Silty and gravelly sands	SM	5.0–9.0	6.0–9.0	High
Clayey and silty sands	SM – SC & SC	9.5–35.0	5.0–7.0	High
Silts and organic silts	ML – OL	23.0–33.0	1.1–14.0	Low to high
Silts and organic silts	ML	33.0–45.0	14.0–25.0	Very high
Clayey silts	ML – CL	60.0–75.0	13.0	Very high
Gravelly and sandy clays	CL	38.0–65.0	7.0–10.0	High
Lean clays	CL	65.0	5.0	High
Lean clays	CL – OL	30.0–70.0	4.0	High
Fat clays	CH	60.0	0.8	Very low

Source: Moulton, L.K., Highway subsurface design, FHWA-TS-80–224, U.S. Department of Transportation, Washington, DC, 1980.

REFERENCES

Barber, E.S. and Sawyer, C.L. 1952. Highway subdrainage. *Proceedings, Highway Research Board*, pp. 643–666. Washington, DC: Highway Research Board.

Ridgeway, H.H. 1976. *Infiltration of Water Through the Pavement Surface, Transportation Research Record No. 616*, pp. 98–100. Washington, DC: Transportation Research Board.

4 Pavement Drainage Structures

4.1 SOURCES OF WATER

Generally, there are a number of sources of water in a pavement. These consists of groundwater (water table), water from artesian aquifers, snow melt water, as well as precipitation coming from the surface through joints and cracks (Figure 4.1). Furthermore, there can be flood water caused by intense rain or flooding in nearby water bodies such as rivers. The flood water can be stagnant or flowing, with varying levels of velocity. High velocity flowing water can lead to erosion and washouts of pavement subsurface layers and complete destruction of the entire pavement structure. However, such occurrences are relatively rare, only caused by natural calamities. The more persistent and common problems are those that are caused by ground water or precipitation water entering through the surface. In pavements situated in colder climates, the freeze-thaw effect is also quite common and has to be regularly taken into consideration during the pavement design.

4.2 TYPICAL DRAINAGE STRUCTURES IN PAVEMENTS

Typical drainage structures (Figure 4.2) in roadways can be divided into surface and subsurface drainage. These structures are provided for collection of water and/or conveyance and discharge.

For surface drainage, pavements are provided with slopes to facilitate the quick conveyance of water to side drains or ditches, and hence to minimize the amount of time the water is present on the travelled lanes. Once the water reaches the side of the pavements, it enters a ditch or an inlet, which lets it fall to a storm drainage system, consisting of an open or a closed channel, which runs longitudinally along the pavement. This drainage system then discharges the water to a larger body of flowing water such as a stream. However, there can be intermediate stages where the conveyed water may need to be detained or retained in order to avoid transmitting pollutants to the stream to which it is finally discharged.

For subsurface drainage system, the components consist of a drainage layer underneath the surface layer with the proper slope to drain the water to the sides, and a side drainage system consisting of longitudinal perforated pipes that are encased in stones and/or wrapped with geotextiles. These side drainage pipes are usually embedded in

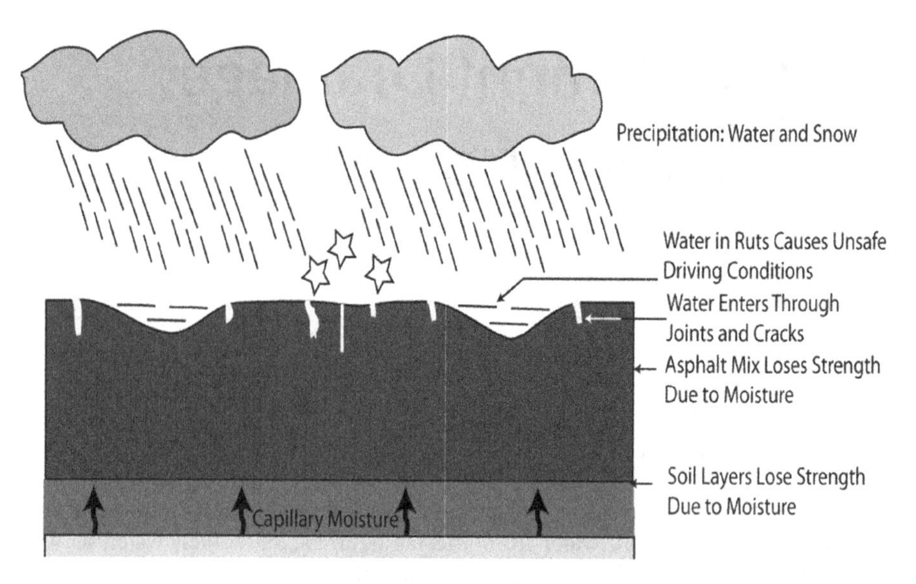

FIGURE 4.1 Sources of water in pavements.

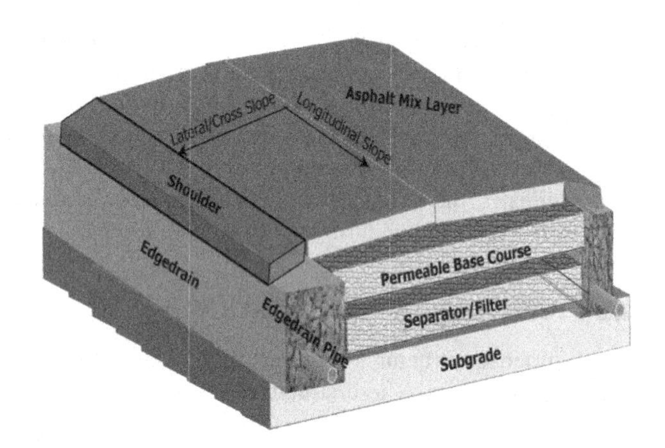

FIGURE 4.2 Surface slope and subsurface drainage layers.

the subgrade with the encasement running up to the level of the drainage layer. The drainage layer could be a layer of aggregates with high permeability or a geotextile-geodrain system in which a relatively thin but extremely permeable layer is created between two sheets of a geotextile. An aggregate drainage layer is generally protected from the underlying layer of finer aggregates with a filter layer.

Culverts also constitute an important part of drainage system. Generally, a culvert is provided to convey flowing water from one side of a road to the other through a drainage channel, in such a way that the road is not affected significantly during high water levels, and the water flow is safely and efficiently maintained – sometimes the water flow may be necessary to ensure unrestricted movement of wildlife/fish also. Note that a culvert is used only when the flowing water body is relatively narrow – beyond a certain width, such as for a river, a bridge will be necessary.

4.3 DRAINAGE STRUCTURES IN AIRPORTS

In airports, the avoidance of standing water is a critical necessity – since standing water generally attracts wildlife including birds, which pose a severe threat to the safe operation of aircrafts. Also, the drainage system should be designed in such a way that the discharged water does not cause any adverse impact on the environment. The U.S. Federal Aviation Administration (FAA) maintains a standard for the design of drainage structures in airports. The relevant FAA code for design of drainage structures in airports is 150/5320-5D (Date: 8/15/2013) (www.faa.gov/documentLibrary/media/Advisory_Circular/150_5320_5d.pdf). The drainage system consists of components for collection of storm water, and its conveyance and discharge. The surface drainage is dictated by transverse and longitudinal slope, pavement roughness and inlet spacing and capacity. For conveyance, the channel or ditch should be designed to prevent erosion and the efficient conveyance of the collected water, without causing surface ponding or flooding. For the discharge system, consideration needs to be made regarding storm water quality and quantity, which are dictated by local authorities, and which may require the use of detention basins or other best management practices (BMPs).

Prior to the design of a drainage system for pavements, a general investigation should be made of the area and the runoff contributing area (tributary area) to ascertain the following:

- Capacity, elevations and condition of existing drains.
- Topography, size and shape of drainage area, and extent and type of development.
- Profiles, cross sections and roughness data on pertinent existing streams and watercourses; and location of possible ponding areas
- Climatic conditions and precipitation characteristics.
- Soil conditions, including types, permeability, vegetative cover, depth to and movement of subsurface water, and depth of frost.
- Outfall and downstream flow conditions, including high-water occurrences and frequencies.
- Effect of base drainage construction on local interests, facilities and local requirements that will affect the design of the drainage works.

In addition, the following considerations also need to be made:

- For diversion of runoff, efforts needed to avoid undesirable downstream conditions.
- Agreements needed to obtain drainage easement or avoid interference with water rights.
- Evaluation of possible adverse effects of disposal of drainage water on water supply system.
- Appropriate policies to protect and restore the environment (for example National Environmental Protection Agency, NEPA, Public Law [PL] 91–190).

- Federal guidelines to "maintain compatibility and minimize interference with existing drainage patterns, control flooding of the pavement surface for design flood events, and minimize potential environmental impacts from facility-related storm water runoff". The planning for storm water drainage system should begin in the early planning phase of any transportation project, and sustainability considerations given in the Whole Building Design Guide (WBDG, www.wbdg.org/), should be consulted with.
- Federal, state and local laws that can impact storm water drainage design.

In general the drainage structures that are used in conventional pavements are applicable for airfields. However, these structures must be capable of withstanding the load coming from the heaviest aircraft in operation, or future aircrafts. For heavy loads, three gear configurations are used: Type A – Bicycle; Type B – Tricycle; and Type C – Tricycle (FAA AC 150/5320–6) (Figure 4.3).

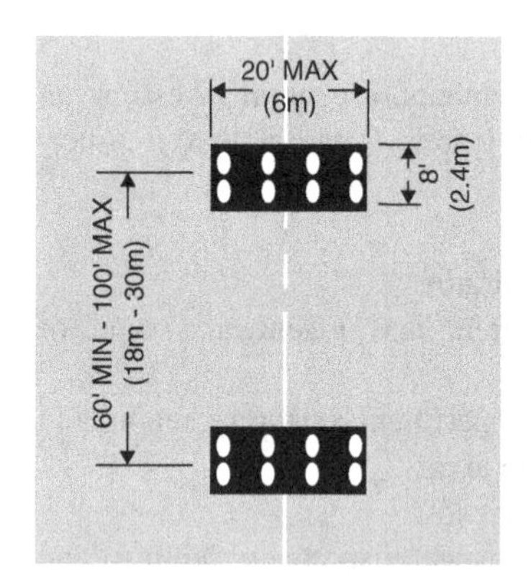

Type A - Bicycle Gear Configuration

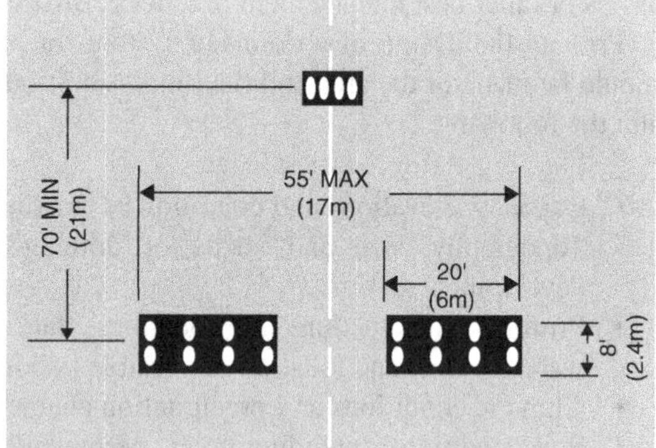

Type B - Tricycle Gear Configuration

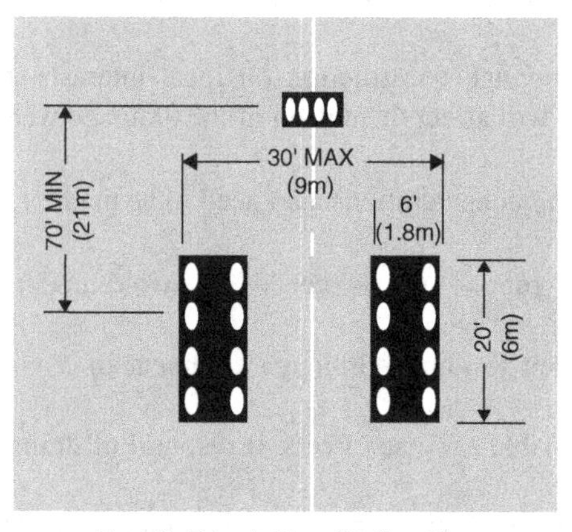

Type C - Tricycle Gear Configuration

FIGURE 4.3 Different types of gear configuration for aircrafts. (Source: FAA Circular, 150/5320-5D; Date: 8/15/2013.)

For a given aircraft gross load, each configuration is used in the design of each drainage component, and the configuration corresponding to the most conservative design is selected as the design gear configuration for that component. Each gear configuration has two wheel groups of eight wheels each. The area occupied by each group of wheels is 20 feet by 6 feet or 8 feet, and supports one-half of the gross weight. The wheels are laid out at uniform spacing within each wheel group.

The Type A – Bicycle configuration consists of two wheel groups located along a single line parallel to the primary aircraft axis (i.e., parallel to the line of travel), but with the major axis of each wheel group oriented perpendicular to the primary aircraft axis. The Type B – Tricycle configuration includes a nose gear and has wheel groups whose major axes are coincident and perpendicular to the major aircraft axis. The Type C – Tricycle configuration includes a nose gear and has wheel groups whose major axes are parallel to, and equidistant from, the principal aircraft axis.

Those structures that are supposed to take direct load should also consider a maximum breaking load of 0.7 g (for no-slip brakes). Direct loads are applicable for decks and covers, and includes structures such as manhole covers, inlet grates, utility tunnel roofs and bridges. Manhole covers should be designed for 100-kip wheel loads with a tire pressure of 250 psi. For structures, the following specifications are given:

1. For spans of 2 feet or less in the least direction, apply a uniform live load of 250 lb/in^2.
2. For spans greater than 2 feet in the least direction, the design will be based on the number of wheels that will fit the span. Wheel loads of 50 to 75 kip should be considered.

It is specified that for diagonal taxiways or apron taxi routes which are required to support both in-line and directional traffic, load transfer at expansion joints are not to be considered except in cases where there is specific knowledge about the long-term load transfer characteristics of a particular feature that supports the use of load transfer in the design of a particular drainage structure.

4.3.1 Detention and Retention Ponds

An increase in paved area due to the construction of a pavement can lead to an increase in the peak discharge due to storm water runoff. Methods of controlling the quantity and quality of storm water runoff include temporary storage areas known as detention or retention facilities, which can reduce the frequency and extent of downstream flooding, soil erosion, sedimentation and water pollution, and reduce the required size of drains and hence facilitate savings in the storm drainage systems. Detention (holds water for shorter period of time, hours to days) and retention (holds water for relatively longer periods of time, months) facilities help reduce the peak discharge and volume from a watershed, and bring it closer to that in the pre-construction stage.

The safety of the detention/retention facilities must be considered by providing adequate measures to prevent access to people and other hazards, and emergency escape means. The inlet and outlet pipes, specifically that are connected to underground storm drains should be provided with removable, hydraulically-efficient grates and bars, and the pond should be fenced. If the detention basin is to be made part of an active recreation area, then mild slopes should be provided on the sides, and the basin should be located away from busy streets and intersections. The outflow structures should be designed in such a way that the flow velocities are not high enough to force a person, or pin down a person, to outflow under current. Finally, for airfields, proper consideration should be given to the attraction of wildlife, particularly waterfowls, to such waterbodies, which pose significant threat to aircrafts (AFPAM 91–212 or FAA AC 150/5200-33B). Seventy eight percent of the strikes have been reported to be under 1,000 ft, and 90% under 3,000 ft. The FAA recommends three different zones for separation distances for any wildlife (birds, mammals, reptiles – avoided, eliminated or mitigated) that can pose hazard to aircrafts. These zones are 5,000 ft from the Aircraft Operations Area (AOA) for piston powered aircraft (only) airports, 10,000 ft for airports serving turbine-powered aircrafts and a 5-mile range from the farthest edge of the AOA to protect the approach, departure and circling area.

The FAA recommends draining of detention ponds within 48 hours and leaving them completely dry between rainfalls. If continuous water flow is expected, then to prevent nesting habitat of wild animals, the pond should be provided with concrete or paved pad and/or ditch/swale in the bottom to prevent vegetation. If the pond cannot be drained quickly then it should be covered with physical barriers such as bird balls (Figure 4.4) (also used in drainage ditches), wire grids, pillows or nettings, after ensuring that they will not adversely affect water rescue.

For new storm water drainage systems within the perimeters suggested above, FAA recommends designs that would not result in above-ground standing water and the use of underground water infiltrations systems such as French drains and buried rock fields, using steep-sided, rip-rap lined, narrow, linearly shaped water detention basins and elimination of all vegetation around detention basis.

4.4 CULVERT

A culvert (Figure 4.5) (FHWA, 2012) is a conduit to convey flow of water through a roadway embankment. The variables associated with a culvert are materials, shapes and configuration. The important factors affecting the design of culvert are roadway profile, channel characteristics, expected flood damage, service life and costs of construction and maintenance.

Culverts are needed when a bridge is not required from hydraulic consideration and the consideration of ice and debris potential. Also, it may be used, when it is more economical than bridge. If both can be used, a cost comparison is necessary. The initial cost is less than that of a bridge because of the smaller opening compared to a bridge. The culvert is designed to accommodate a headwater whereas the bridge is designed for freeboard at the design discharge. The cost should be compared against the cost of damages caused by flooding and that of providing larger openings and natural invert that may be required under special design considerations such as aquatic organism passage (AOP). Also, long span culverts (such as of length >20') are considered as bridges and may be subjected to same design, construction and maintenance requirements.

The most common culvert shapes are circular, box (rectangular), elliptical and pipe arch for manufactured culverts, whereas embedded culverts may consist of box and arch shapes (Figure 4.5). The factors affecting the selection of the shape of

FIGURE 4.4 Use of Bird Ball for covering open water areas in airports. (Courtesy, Bird-X, Inc.)

the culvert include cost, upstream water surface limitations, roadway embankment height and hydraulic performance.

Commonly used modern culvert materials are reinforced and plain concrete, corrugated aluminum or steel, plastic (HDPE) or polyvinylchloride (PVC). The material for the culvert should be selected on the basis of structural strength, hydraulic roughness, durability (corrosion and abrasion resistance) and constructability. Liners may be used inside culverts to protect from them abrasion and corrosion and reduce hydraulic resistance (for example, corrugated metal culverts lined with asphalt mix or polymer).

The selected type of inlet is dependent on hydraulic performance, structural stability, aesthetics, erosion control and fill retention. The different configurations include projecting culvert barrels, cast in place concrete headwalls, precast or prefabricated end sections and culvert ends mitered to conform to the fill slope.

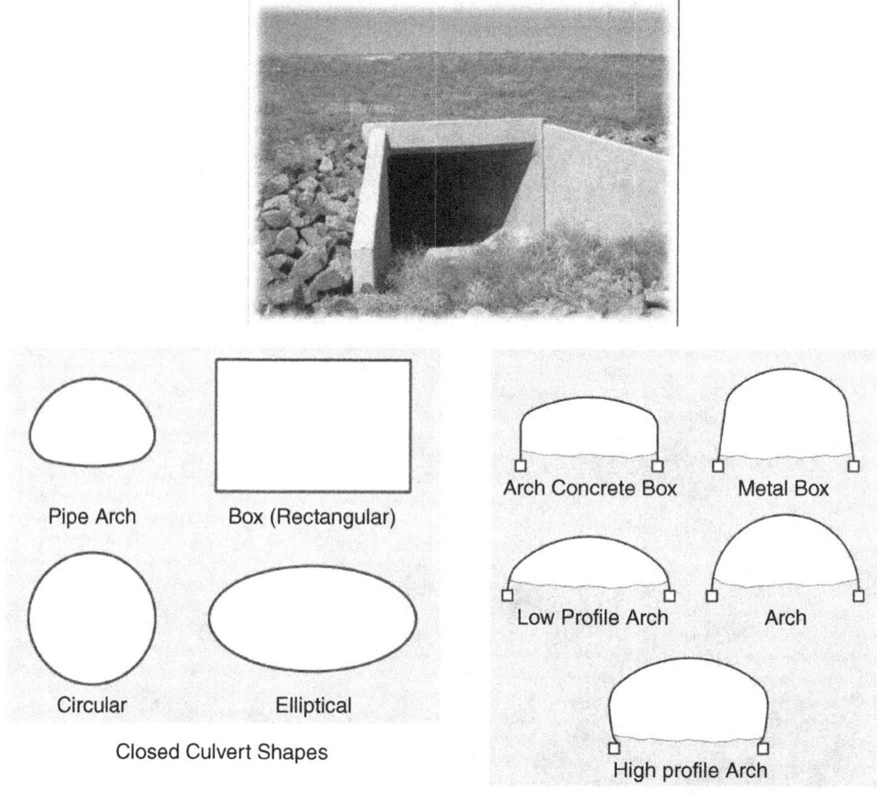

FIGURE 4.5 Culvert and commonly used culvert shapes. (Source: FHWA, Publication No. FHWA-HIF-12–026, Hydraulic Design Series Number 5, *Hydraulic Design of Highway Culverts*, Third Edition, April 2012.)

The inlet section (Figure 4.6) affects the hydraulic capacity of the culvert. Beveled edges are more efficient than square edges because the more gradual flow transition reduces the energy loss and makes it more hydraulically efficient. Even better are side tapered and slope tapered inlets. Slope tapered inlets with depression can increase head on the flow control section and increase culvert efficiency. The provision of tapered inlet will require additional cost but can help reduce the size of the culvert barrel and reduce cost.

This is especially relevant for situations with long culverts or steep slope, existing culvert, which is found to be undersized after a road-widening project or rehabilitation of a culvert under a high fill.

A natural invert is provided by embedded and open bottom culverts, which is desirable in channels with high sedimentation transport consisting of mostly coarse materials such as gravels and cobbles, aesthetics and AOP issues. Generally open bottom culverts are box or arch shape on a vertical wall foundation, which should be designed against scouring potential.

FIGURE 4.6 Different types of inlets. (Source: FHWA, Publication No. FHWA-HIF-12–026, Hydraulic Design Series Number 5, *Hydraulic Design of Highway Culverts*, Third Edition, April 2012.)

The foundation can be protected with riprap. For culverts with high scouring potential, an embedded culvert can be considered. Generally an embedded culvert is circular, box or pipe arch type that is buried in the ground 20–40% of its height.

The advantage of an embedded culvert (Figure 4.7) is that it provides grade control and protection against extreme scour better than that from an open culvert. Latest sources of information for open bottom and embedded culverts is NCHRP Project 15–24, NCHRP 2011 (www.trb.org/Publications/Blurbs/168265.aspx).

In low-water crossing culvert, the water is expected to overtop the road at high flows and cause a stoppage of traffic. This type of culver is less costly and can be considered for low volume roads, when the expected duration of submerge is only few hours or few days in a year. Design details are available from FHWA (2008) and USFS (2006). The low water crossing culvert can be vented or unvented, depending on whether it has or does not have a hydraulic opening beneath the road. An unvented culvert can be considered in arid climates where the channel is dry most of the times in the year. A vented culvert acts as a culvert for flows up to the low design discharge and as a broad crested weir with flow for greater discharge. Signs should be posted on low water crossing culverts to warn traffic of danger of currents that can be strong enough to sweep off vehicles. Generally rock riprap along embankments or paved

FIGURE 4.7 Example of embedded culvert. (Source: FHWA, Publication No. FHWA-HIF-12–026, Hydraulic Design Series Number 5, *Hydraulic Design of Highway Culverts*, Third Edition, April 2012.)

embankments are provided to restrict scour or erosion of the structure due to the flowing water, especially during overtopping.

Long span culverts (20–40', 7–14 m) are generally provided on the basis of structural design rather than hydraulic considerations (AASHTO 2002). They may be necessary when their lengths exceed maximum size of pipes, pipe arches and arches or where the slope requires long radius of curvature in the crown or side plates. Such special shapes include vertical and horizontal ellipse, underpasses and low and high profile arches. Unique concerns and features of long span culverts include:

1. Proper bedding, selection and compaction of backfill.
2. Avoiding unbalanced loads during backfilling of multiple barrels.
3. Anchorage of ends to prevent damage at inlet.
4. Use of normal open channel flow calculations for design since there is less chance of overtopping.

The functions of a culvert can be listed as follows:

1. Provide cross damage for a stream channel
2. Provide flood pain relief
3. Provide relief for road drainage
4. Act as outlet control structures for detention ponds
5. Provide a crossing structure for human or animal traffic
6. Facilitate wildlife movement

4.4.1 Culvert Hydraulics

Culvert design is complex because of a number of possible conditions that may be present which include depth of flow in the barrel (partial or full), barrel features and inlet geometry. The approach followed by FHWA (HEC5) (Figure 4.8) considers three types of flow that are found in a culvert that is flowing under partly full condition. The upstream channel experiences subcritical flow, and the culvert barrel has supercritical flow, with critical depth occurring at the culvert inlet.

The location of the control section in a culver defines the type of control that is expected in it – inlet or outlet, for which the combination of factors affecting the hydraulic capacity of the culvert are different. The principal factor governing the type of flow is the slope of culvert, also known as the slope of the barrel. If the culvert barrel can flow more water than the inlet can accept, then the culvert is said to be under inlet control (Figure 4.8), with the control section occurring just inside the entrance. This is also the location for the critical depth, and the flow downstream of this point is supercritical. The flow is controlled by the upstream water surface elevation and the inlet geometry – shape, cross sectional area, configuration. If the barrel cannot flow as much water as permitted by the inlet opening then the culvert is operating under outlet control, and in this case the control section is at the barrel exit or downstream of it (Figure 4.9). The governing factors are those controlling inlet control case, water surface elevation at the outlet and the characteristics of the barrel.

The energy that is required for flow though the barrel is provided by the increased upstream water surface elevation. The depth of the upstream water measured at the culvert inlet is known as the headwater depth, whereas the depth of water measured at the outlet is known as tailwater depth. Tailwater controls the flow of culvert under outlet control conditions – a high tailwater can make a culvert flow change to outlet control from inlet control, and is itself dictated by the characteristics of the downstream channel at the given design discharge. High depth of tailwater can be caused by downstream obstructions such as another channel or another bridge or culvert. If flow velocity in the culvert is higher than that in the channel, because of narrowing of the channel, scour of the streambed and erosion of the sides can be expected near the

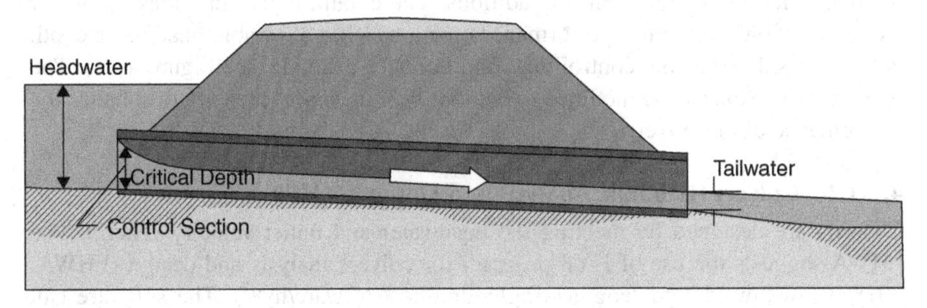

FIGURE 4.8 Flow in a partly full culvert. (Source: FHWA, Publication No. FHWA-HIF-12–026, Hydraulic Design Series Number 5, *Hydraulic Design of Highway Culverts*, Third Edition, April 2012.)

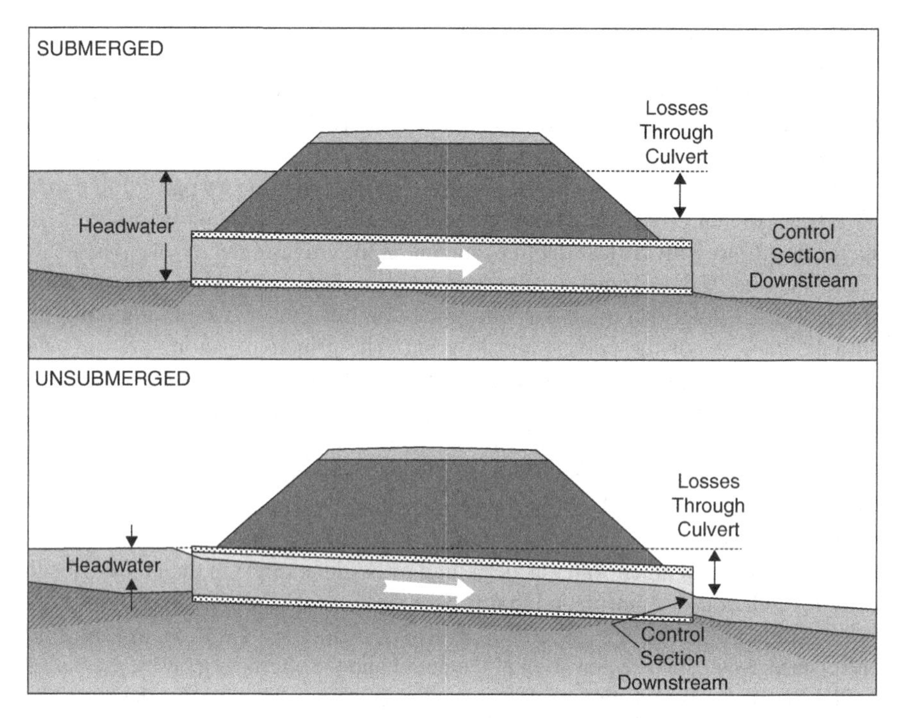

FIGURE 4.9 Outlet control flow. (Source: FHWA, Publication No. FHWA-HIF-12–026, Hydraulic Design Series Number 5, *Hydraulic Design of Highway Culverts*, Third Edition, April 2012.)

culvert outlet. Steps to avoid these problems include increasing the barrel roughness, and the use of energy dissipation and outlet protection devices. For partially filled culverts operating under inlet control, outlet velocities can be reduced by reducing the barrel slope and adding a rough section.

The basis of evaluation of hydraulic capacity and design of a culvert is Performance Curve, which is a plot of discharge (x) versus headwater depth (y), generally plotted for both inlet and outlet control conditions. The condition that provides the lower amount of flow (minimum performance) for a specific allowable headwater depth is then considered as the controlling condition (for example, see Figure 4.10). The inlet configuration can be modified to better utilize the barrel capacity to enhance the performance of the culvert.

4.4.1.1 Culvert Hydraulic Analysis and Design by HY8

Culverts are designed by meeting the headwater and outlet velocity criteria. The FHWA suggests the use of HY8 program for culvert analysis and design (FHWA, 2018, www.fhwa.dot.gov/engineering/hydraulics/software/hy8/). The software can be downloaded from fhwa.dot.gov/engineering/hydraulics/software/hy8 (version 7.5, from 2016). The software also enables the analysis of special culvert geometrics, rehabilitation timings and analysis of energy dissipation strategies to mitigate scour for high velocity outlets. HY8 cannot analyze unsteady flow conditions. In such

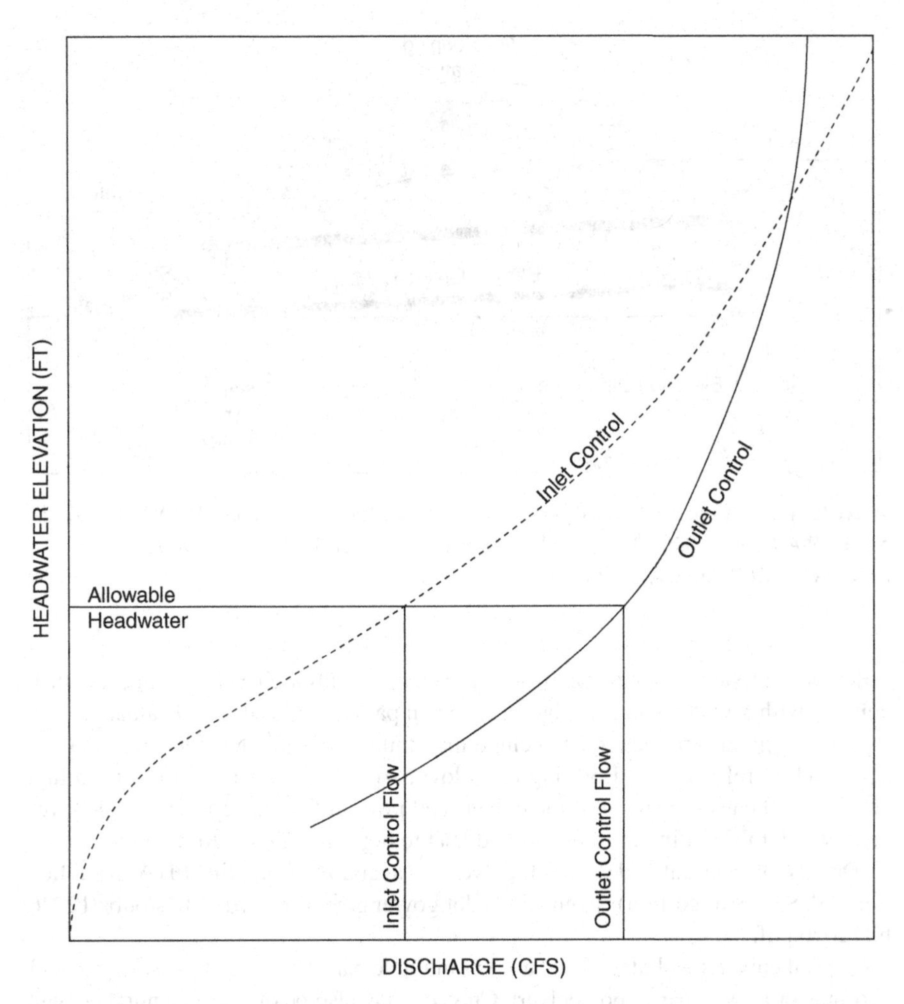

FIGURE 4.10 Culvert Performance Curve. (Source: FHWA, Publication No. FHWA-HIF-12–026, Hydraulic Design Series Number 5, *Hydraulic Design of Highway Culverts*, Third Edition, April 2012.)

cases, water surface profile model (such as HEC-RAS) may more fully consider upstream headwater and downstream tailwater conditions. Figure 4.11 shows the commonly used terms that are used in analyses of culverts.

Hydraulic performance of a culvert is evaluated in terms of factors such as headwater elevation and outlet velocity, for which HY8 can be utilized. The limiting combination of tailwater, barrel characteristics and inlet characteristics are used to determine headwater elevation.

The hydrological considerations include return period (commonly taken as 50 years) and checking flow (larger than the design return period flow). Usually peak

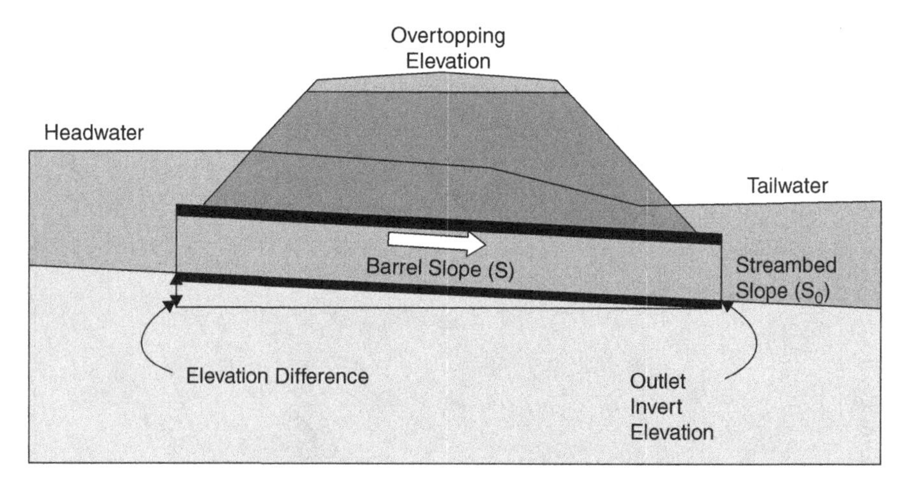

FIGURE 4.11 Commonly used terms in culvert hydraulics. (Source: FHWA, Publication No. FHWA-HIF-12–026, Hydraulic Design Series Number 5, *Hydraulic Design of Highway Culverts*, Third Edition, April 2012.)

flow rate and steady state conditions are assumed although velocity and depth in culverts with low flows in association with fish passage may also be evaluated.

The design criteria generally include maximum allowable headwater, outlet velocity and barrel velocity. Both high and low barrel velocities are harmful – a high velocity will cause abrasion of the culvert and prevent fish passage while a low velocity will inhibit sediment transport and lead to deposits in the barrel.

Details of hydraulic design of culvert is available from the FHWA publication HDS5 (3rd edition) from fhwa.dot.gov/engineering/hydraulics/pubs/12026/hif12026.pdf.

Typical culverts include single barrel or multiple barrel outlets, such as a single culvert pipe or a triple barrel box culvert. Culverts may also be of multiple numbers with different size, shape, material or slope, such as a four barrel box culvert along with a double barrel pipe culvert at a high elevation with a different slope. Analysis can also be conducted with HY8 for aquatic organism passage design and open bottom culverts.

Crossing data includes discharge, tailwater and roadway data. Culvert data includes shape, material, span, size, embedment depth, Manning's n, inlet type, inlet flow condition, inlet depression, inlet station, inlet elevation, outlet station, outlet elevation and number of barrels.

The output contains various parameters and a Performance Curve which plots total discharge versus headwater elevation for inlet control elevation and outlet control elevation.

The process of culvert design consists of the determination of a design flow and elevation (headwater), selection of barrel type and trial size, calculation of headwater (with HY8), and checking whether calculated headwater is less than or equal to the design elevation (headwater). If yes, the design is complete, if not, a different barrel type/size is selected. The process is repeated until a cost effective culvert is selected that meets the design headwater criteria.

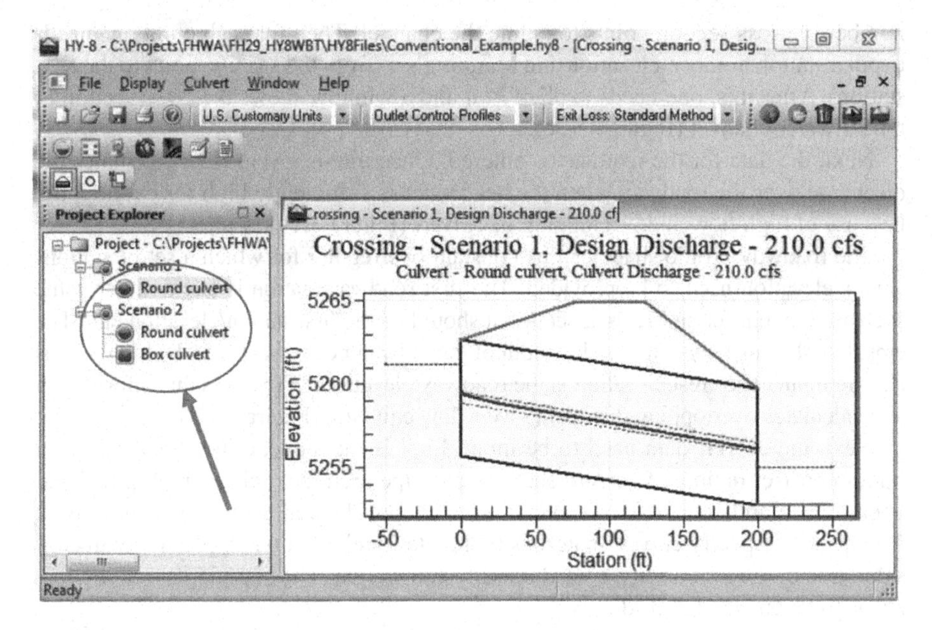

FIGURE 4.12 Different shapes of culverts in HY8.

In HY8 the user opens up the main window which gives the option to select a new project or open an existing one (Figure 4.13a). For a new project the data input screen will be presented, whereas for an existing project the data for the culvert selected on the right side of the screen is presented. The crossing properties that need to be input consists of discharge data, tailwater data and roadway data. For culvert, the data include culvert and site data. For evaluating multiple culverts at the same crossing, one can use the *Add culvert* feature and the software will consider the contribution of each culvert in the analyses (Figure 4.13b).

For example, consider roadway width = 49', elevation = 5307.7'; culvert inlet at 5300.7'; outlet elevation = 5300'. Culvert length = 200'. Requirement: headwater not greater than 1' below the overtopping elevation, for example, maximum allowable headwater elevation = 5307.7-1 = 5306.7'.

Put the discharge method (can use other methods by the menu for discharge method) as minimum, design, maximum and enter the values as 100, 183 and 200 cfgs respectively (Figure 4.13c).

Enter the tailwater data: put channel type, bottom width, side slope (H:V), channel slope, Manning's n (channel) and channel invert elevation. Input: trapezoidal, 5.00, 3.00, 0.0034, 0.0550, 5300.00. Then click on *View* to see the table for the rating curve, which lists flow, elevation, depth, velocity, shear (Figure 4.13d). There is an option to input other channel cross sections, constant tailwater or user supplied rating curve. There is also an option to input natural channel cross section by selecting irregular channel.

Then there is a need to *define it*; in the next window there is the option to import a cross section that has been previously generated, or enter data in a new window. For defining a new one, one needs to put a slope of the tailwater channel as well as the number of cross section points to define the channel. Then for each point one needs to put a station number, elevation and Manning's n (from the specific point to the next station). After this, one needs to click *Plot* and a window shows the cross section of the tailwater channel (Figure 4.13e).

Next, the data for the roadway is entered. Once that is given, HY8 calculates the discharge over the roadway when the headwater is sufficiently high for overtopping because of high flow and/or clogging of culvert (Figure 4.14).

The roadway profile shape can be constant or irregular for which a set of stations versus elevation need to be provided. The first roadway station is any number value such as zero, but for an irregular section it should be the first station/elevation pair. The crest length is usually taken as the width of the tailwater cross section. The top width is also an input. The crest elevation is the roadway elevation. HY8 takes this information and calculates overtopping flow using weir flow equation (Figure 4.13f).

Next, the culvert data need to be input. First is the slope, either standard, open bottom or user defined. Standard sections are pipe arch, box, circular, elliptical and open broken arch, concrete box, metal box, low profile arch, arch, high profile arch. One needs to specify culvert materials (concrete, steel, aluminum, plastic) with consideration of the availability, cost, durability and weight. The Manning's n coefficient needs to be changed accordingly.

Next enter a name, the span, rise, embedment depth and Manning's n (which is suggested), culvert type and inlet configuration and inlet depression. For example:

Name: single box, concrete box, concrete, 5.0, 4.0', 0, 0.0130, straight, 1:1 barrel (45° flow).

Next indicate how the culvert fits with the embankment by filling in the site data, by using either the culvert invert data (more common) or the embankment toe data. Specify the inlet station and the elevation and outlet station and elevation (outlet station – inlet station = length of the culvert) and the number of barrels.

A culvert in HY8 is defined as consisting of one or more barrels. Each barrel has been previously specified in the culvert section. In the embankment toe data option, specify the station and elevation of the embankment toe on the upstream and downstream sides of the embankment and the embankment slope which lets HY8 match the culvert inverts to match toe data.

Next select *Analyze* to initiate calculations, which brings out summary tables and water surface profiles of the conditions at inlet limit flow through the culvert; at low flows the inlet is unsubmerged, flow travel through barrel and the outlet is also unsubmerged. As flow increases, inlet continues to limit or control flow, causing an increase in headwater. As the headwater increases the depth of flow in barrel and tailwater increase but not enough to cause the barrel to flow full or for the tailwater to control the flow (Figure 4.13g).

If the flow is limited by the condition in the barrel or the outlet, then as the flow increases, the resistance of the barrel limits flow, the barrel fills up and the inlet is submerged. As the flow increases, tail water increases and causes the outlet to submerge, causing the barrel to flow full and headwater to rise further. HY8 computes

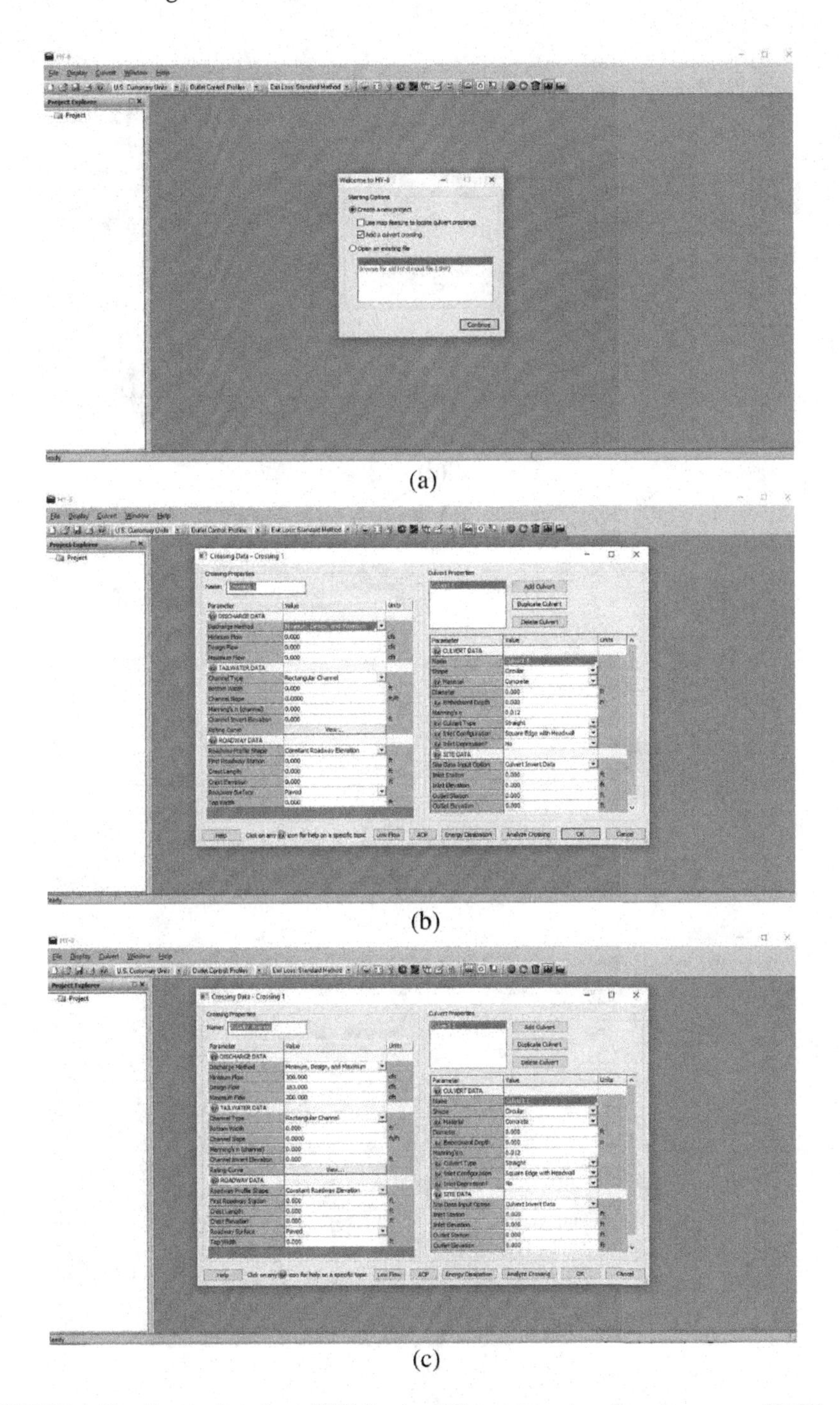

FIGURE 4.13 Screenshots from HY8 for the different steps (continued on pages 60–61).

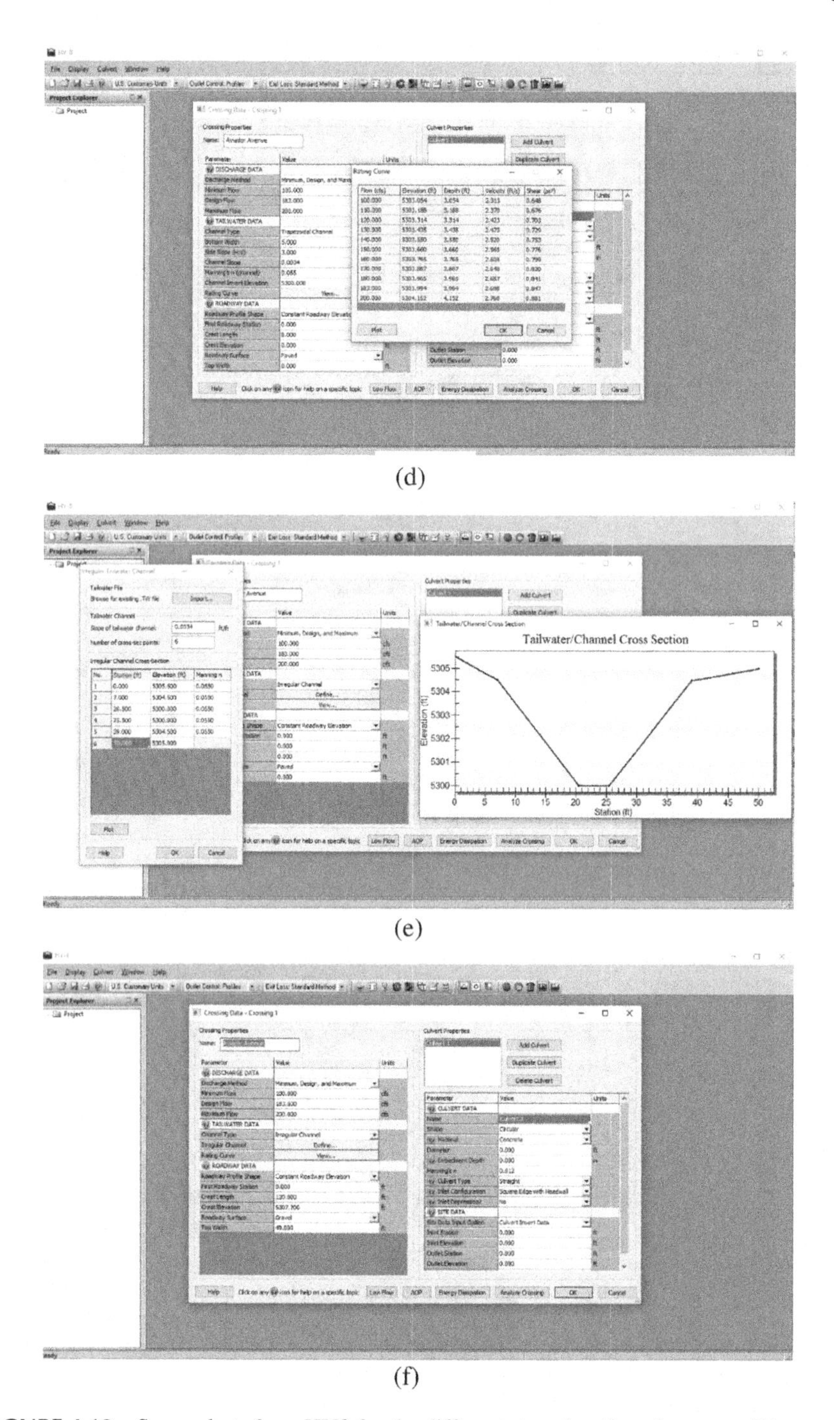

FIGURE 4.13 Screenshots from HY8 for the different steps (continued on page 61).

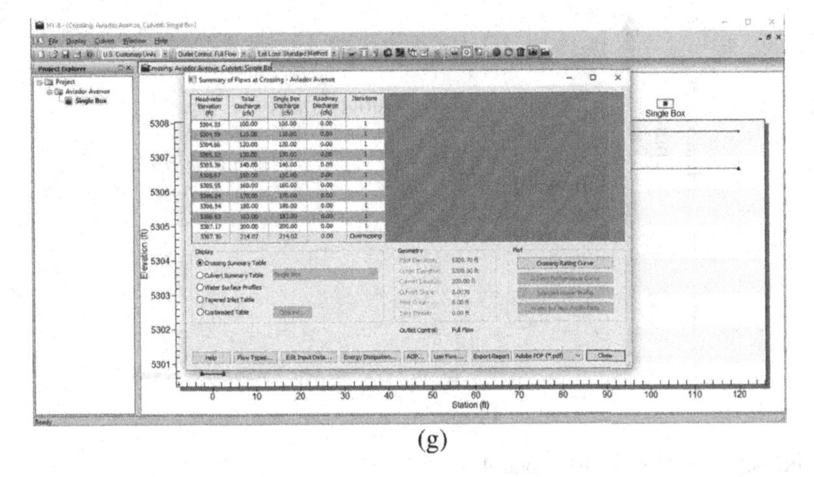

(g)

FIGURE 4.13 Screenshots from HY8 for the different steps.

FIGURE 4.14 Overtopping of culvert. (Courtesy: Alexander Mann, Maine DOT.)

both flows and takes the one which gives the largest headwater at a given flow from the two sets of data. It may be so that the process is inlet controlled up to a point and outlet controlled at higher flows (Figure 4.15).

HY8 provides the crossing summary table, showing the headwater elevation, total discharge, culvert flow, roadway discharge and how many iterations are

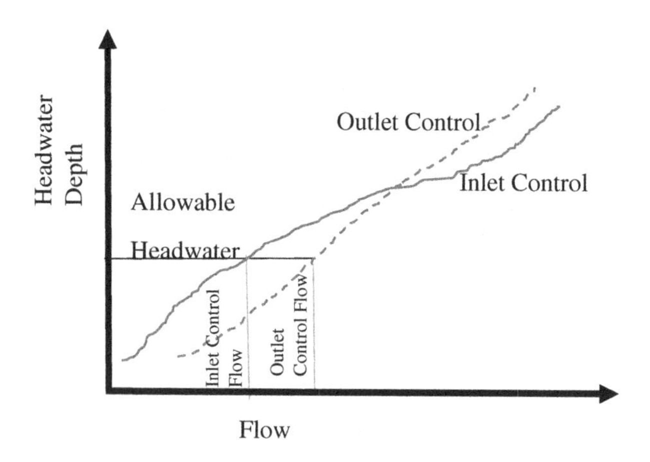

FIGURE 4.15 Inlet and outlet control.

needed to achieve a balanced flow between culvert and roadway to match total discharge. One needs to check whether the headwater is less than equal to the maximum specified headwater at the design discharge. If yes, the design criteria are satisfied.

4.4.2 Repair and Rehabilitation of Culverts

Cost of maintenance for culverts mainly comes from those that are required against erosion at inlet, outlet and the culvert invert, sedimentation, buildup of debris, and the repair of the embankment and road if there is overtopping. The appropriate repair and rehabilitation technique is determined and a decision to replace or repair can be made after a careful consideration of different factors which include both the condition of the culvert and its hydraulic performance. The three common choices are replacement of the lining, detailed repair based on man-entry or replacement. The relevant details are given in the Culvert Assessment and decision-making Procedures Manual (FHWA 2010 https://www.fhwa.dot.gov/engineering/hydraulics/library_arc.cfm?pub_number=208&id=166).

4.4.3 Culvert Economics

The basis for the economic evaluation of culvert installation that is needed for the selection of a culvert is the variability of floods and their probabilities that are expected at the site, and a balance is sought between the initial cost of construction and the damage that could be expected if the culvert is not able to withstand the different expected floods. The annual cost of maintenance of the culvert which includes capital expenses, maintenance cost and damages due to risk of flooding, should be minimized over the life of the roadway in which the culvert is installed. Critical durability issues include abrasion and corrosion, and important considerations, besides initial costs of different types of shapes and sizes, are durability, maintenance cost and replacement cost. The cost of future maintenance and replacement, which will dictate the total annual cost, and not only the initial construction cost, should be ideally considered during the selection process. The major benefits of a culvert are the reduction in the interruption of traffic flow during a flood and increased safety, while the major costs include those that are required for initial construction of the embankment and the culvert, the annual maintenance cost of the culvert and costs due to any damage caused by flooding.

The initial construction cost can be reduced by comparing and allowing bids for appropriate material and shape of culverts with equal hydraulic capacity. The considerations that are required for detailed economic analyses include the structural strength that is required, the slope of the channel and water tightness at joints. The material of the culvert quite often dictates the shape of the culvert, which is also an important consideration. While circular culvert are favored in view of their ability to support high loads, moderate cost and hydraulic efficiency, pipe arch or ellipse, which are more expensive may be necessary at locations with restricted fill heights. The consequences of the selection of a specific shape of the culvert are many. Foundation design and prevention of scour are important considerations for arches, which, however, are beneficial for AOP because of their ability to provide a natural streambed. Benefits of certain types of culverts include quick construction and low transport costs for structural plate conduits, ability to use multiple cells for box culverts and lower construction time for precast box culverts (Figure 4.16).

The service life of the culvert is determined from the consideration of the ability to reline or replace the culvert versus the service life of the roadway. For cases where replacement is easy and the roadway has a long service life, a relatively short service life may be selected. For areas with high potential of flood induced damage a risk analysis could be performed to select the best culvert with the optimum capacity. The total expected cost (TEC) consists of annual capital cost and an annual economic risk cost. The objective of risk analysis is to minimize the TEC (least TEC, or LTEC) with engineering and economic considerations. The difference between traditional design and risk analysis based design can be explained as follows. In traditional designs the specified design frequency flood, or limits on backwater, contain the risk that is associated with the design of a culvert, and the consequence of this risk will be site dependent. However, in risk analysis based design, the impact of each alternate design at specific points in the flood frequency curve is estimated explicitly, and the flood frequency for overtopping of the road is given more importance than the design flood frequency. The alternatives in risk analysis based design are selected on the basis of a number of factors such as minimum design flood criteria, limits imposed by roadway geometrics, legislative mandates regarding flood plains and channel stability.

4.5 DETENTION AND RETENTION PONDS

The purpose of a detention pond is to limit the peak outflow rate of storm water in a watershed to a level that existed prior to the development of the area (such as construction of pavement) (Figure 4.17).

The various forms of detention ponds consist of impoundments, collection and conveyance facilities and underground tanks, and majority are dry ponds as opposed

Structural Plate Culvert

Multi-cell Box Culvert

FIGURE 4.16 Structural plate and multi-cell box culverts. (Source: FHWA, Publication No. FHWA-HIF-12–026, Hydraulic Design Series Number 5, *Hydraulic Design of Highway Culverts*, Third Edition, April 2012.)

Detention Pond

Retention Pond

FIGURE 4.17 Detention and retention ponds. (Source: Laramie County Conservation District. Best Management Practice for Stormwater Runoff)

to wet ponds which maintain a permanent pool of water. While the dry ponds (that discharge water quickly) facilitate settling of larger size sediments (sand and larger silts) and associated pollutants, the wet ponds (which are typically used for retention ponds) are more effective in removing a larger part of the sediments (including clays and silts), through a longer time, biological reaction and by avoiding resuspension of the once deposited sediments.

Detention ponds should be designed with consideration of design rainfall intensity, duration and frequency data, and the outlet structure should be able to limit the maximum outflow to the allowable release rate as dictated by the existing or new runoff rates, downstream channel capacity, potential flooding conditions and local rules. The volume of the required detention pond can be designed with the hydrograph approach (HEC-22). The facility should have an auxiliary outlet to allow flow in the case of excessive inflow or clogging of the main outlet. The auxiliary outlet should be placed such that the overflow passes through a predetermined route such as open channel, swale or approved storage or conveyance structures. The dry detention pond should allow quick removal of water in order to make room for water for subsequent storms, and may need a paved low flood channel to ensure complete removal of water. The facility should have proper access for maintenance.

The purpose of a retention facility or pond is to provide storm water quality control in addition to controlling its quantity. It helps in effective removal of pollutants, and in some cases for ground water recharge and storm water storage. Impoundments, collection and conveyance facilities such as swales or perforated conduits and on-site facilities such as pervious pavements in parking lots or roadways can be used. In addition to meeting design requirements of detention ponds (although the entire runoff is not needed to be removed after each storm), retention ponds should also have sufficient depth and volume for any other desired activity, protected shoreline to prevent erosion, if expected, a provision to lower the elevation of water or draining for cleaning, maintenance or emergency operations and have sufficient safety factors for any dike or earth dam within the structure. For safety, it is advisable to have a safety bench or ledge along the perimeter, with emergent vegetation, that will discourage people from going into the pond and also provide a natural shoreline.

To allow infiltration of water and allow regaining of capacity for a subsequent storm event, the bottom should be pervious, and particulates should be able to settle and get removed from the water before infiltration.

4.5.1 Estimating Discharge for Detention and Retention Ponds

For the design of detention/retention facilities, three considerations must be made: release time, safety and maintenance. The release time of the detention/retention facility will dictate the peak discharge in a downstream main channel, and it should be selected in such a way that although it reduces the runoff rate to an individual storm water conveyance channel, it results in increasing the peak discharge flow of the main downstream channel by letting the discharge reach it at, or near, the same time as its original peak discharge. Such problem, which can lead to downstream flooding, is specifically noted for cases with multiple detention/retention facilities. Maintenance of storm water facilities should be conducted at regular intervals.

Such work involves the following: inspection, mowing, sediment/debris and litter control, nuisance control and structural repair, and replacement. Inspections should be done to ensure proper functioning and no damage, a few months after construction, annually thereafter and also after major rainstorms. Mowing should be done at least twice a year to control the growth of plants. Removal of debris/sediment and trash from ponds should be done at least twice a year, with specific emphasis on avoiding any clogging of outlet structures. Allowance of positive drainage and periodic debris/sediment control will ensure avoiding nuisance such as odor, insects and plants. Periodic replacement, at intervals that are dictated by site specific conditions, should be done for riser structures, standpipes and inlet and outlet devices that are known to deteriorate over time.

4.6 SURFACE DRAINAGE SYSTEM

If the water from rainfall exceeds the depth that is required for hydroplaning for a specific pavement width and rainfall intensity, hydroplaning can be expected. In a broad sense, the potential of hydroplaning depends on vehicle speed, tire condition (pressure and tire tread), pavement micro- and macrotexture, cross-slope and grade, and pavement conditions such as rutting, depression and roughness. Apart from the driver's responsibility to control speed, the following steps could be taken to reduce the potential of hydroplaning and/or prevent accidents:

1. Maximize transverse slope and roughness and use porous mixes (such as open graded friction course).
2. Provide gutter inlet spacing at sufficiently close spacing to minimize spread of water, and maximize interception of gutter flow above superelevation transitions.
3. Provide adequate slopes to reduce pond duration and depth in sag areas.
4. Limit depth and duration of overtopping flow.
5. Provide warning signs in sections identified as problem areas.

The basic principle of surface drainage is to design the pavement surface and the adjacent areas in such a way as to facilitate the quick flow of water falling in the form of rain and/or snow on the surface to the sides, and then drain it away to a nearest point of collection. This involves two major components – providing adequate slope to the pavement surface and drainage channels with sufficient flow capacity along the pavement on either side.

The minimum grade for a gutter is 0.2% for curbed pavements. For sag vertical curves, a minimum of a 0.3% slope should be maintained within 15 m of the level point in the curve. In a very flat terrain, the slopes can be maintained by a rolling profile of the pavement. For one lane, the minimum pavement cross-slope is 0.015 m/m, with an increase of 0.005 m/m for additional lanes. Slopes of up to 2% can be maintained without causing driver discomfort, whereas it may need to be increased above 2% for areas with intense rainfall.

In a multilane highway, if there are three or more lanes inclined at the same direction, it is recommended that each successive pair of lanes from the first two lanes

FIGURE 4.18 Tining of concrete pavement.

from the crown line have an increased slope compared to the first two lanes (by about 0.5%–1%), with a maximum pavement cross-slope of 4%. Note that the allowable water depths on inside lanes are lower because of high-speed traffic on those lanes, and hence sloping of inside lanes toward the median should be done with caution. Pavements also have longitudinal slopes – with a minimum of 0.5% and a maximum of 3%–6%, depending on the topography of the region.

A high level of macrotexture on the pavement is desirable for allowing the rain-water to escape from the tire–pavement interface and reduce the potential of hydro-planing. Tining of PCC pavement (Figure 4.18), while it is still in plastic state, could be done to achieve this. For existing concrete pavements, macrotexture can be improved by grooving and milling. Such grooving in both transverse and longitudinal directions is very effective in enhancing drainage. Use of porous asphalt mix layers, such as open graded friction course, allows rapid drainage of water. It is important that such layers are daylighted at the sides, and are constructed over layers that have been compacted adequately to prevent the ingress of water downward.

The coefficient of permeability of the surface mixture is dependent upon many factors, the more important of which are aggregate gradation and density. If the gradation consists of a relatively higher amount of coarse aggregate particles (compared to fine and filler materials), then the permeability is relatively high. On the other hand, permeability decreases with an increase in density.

Generally, the gradation is decided upon from other considerations, and the only available way for the pavement engineer to lower the permeability of the surface layer is through compaction and hence providing adequate density. Note that because of the effect of the gradation, the desirable density will be different for mixtures with different gradations. Also note that no matter how dense the surface layer is, some water will find its way through it, and hence there must be a way to get rid of this water – through the use of the subsurface drainage system (apart from the fact that the materials must be resistant to the action of water to a certain extent). The permeability of new rigid and asphalt mix pavement can be assumed to be 0.2 and 0.5 in/h, respectively. For detailed data on permeability of asphalt pavements, the reader is requested to see Vardanega, 2014.

There can be different types of drainage channels such as ditches, gutters and culverts. The design of these channels means the design of the cross section. This design is accomplished with the help of Manning's formula (Daugherty and Ingersoll, 1954):

$$Q = \frac{K}{n} S^{1/2} R^{2/3} A \qquad (4.1)$$

Where,

Q = the pipe flow capacity, m^3/s
S = the slope of the pipe invert, m/m
n = the pipe coefficient of roughness (0.012 for smooth pipe, and 0.024 for corrugated pipe; FHWA, 1992)
A = the pipe cross-sectional area, m^2
K = 1
R = A/P
P = the wetted perimeter of pipe, m

The practical considerations for design include the elevation with respect to the subgrade (same or lower level than the subgrade), low construction and maintenance costs, and safeguard against slope failure (slope of 2:1 or less).

Curbs are concrete or asphalt mix structures provided along the side of the low-speed urban highways as well as bridge decks to facilitate collection of drained surface water from the pavement surface and protect pavement sides from erosion. Curbs are generally placed with gutters. In rural areas, roadside and median channels are provided instead of curbs and gutters.

For pavement drainage systems, two parameters are selected (Table 4.1), depending on the type of the pavement, design frequency and spread (accumulated flow in and next to the roadway gutter, which can cause interruption to traffic flow, splash, or hydroplaning problems). The spread (T) is constant for a specific design frequency – for higher magnitude storms, the spread can be allowed to utilize most of the pavement as an open channel.

The nomograph shown in Figure 4.19 can be used to design gutter sections. An example is shown on the plot. Manning's coefficients are provided in Table 4.2.

The next nomograph, Figure 4.20 (ratio of frontal to gutter flow), can be used to calculate the frontal flow for grate inlets and flow in a composite gutter section with width (W) less than the total spread (T), which can also be determined from Figure 4.21.

TABLE 4.1
Recommended Parameters for Drainage Systems

Road Classification		Design Frequency	Design Spread
High volume	<72 km/h	10 year	Shoulder + 0.9 m
	>72 km/h	10 year	Shoulder
	sag point	50 year	Shoulder + 0.9 m
Collector	<72 km/h	10 year	1/2 driving lane
	>72 km/h	10 year	Shoulder
	sag point	10 year	1/2 driving lane
Local streets	low ADT	5 year	1/2 driving lane
	high ADT	10 year	1/2 driving lane
	sag point	10 year	1/2 driving lane))

Note: This criteria applies to shoulder widths of 1.8 m or greater. Where shoulder widths are less than 1.8 m, a minimum design spread of 1.8 m should be considered.

Source: From American Association of State and Highway Transportation Officials (AASHTO), *MDM-SI-2, Model Drainage Manual*, 2000 Metric Edition, AASHTO, Washington, DC, 2000.

Example 4.1: Find spread, given flow

Step 1. Given: $S = 0.01$, $S_x = 0.02$, $S_w = 0.06$, $W = 0.6$ m, $n = 0.016$, and $Q = 0.057$ m^3/s trial value of $Q_s = 0.020$ m^3/s.

Step 2. Calculate gutter flow: $Q_w = Q - Q_s = 0.057 - 0.020 = 0.037$.

Step 3. Calculate ratios of (E_o=) $Q_w/Q = 0.037/0.057 = 0.65$, and $S_w/S_x = 0.06/0.02 = 3$. And, using Figure 4.20, find an appropriate value of W/T; $W/T = 0.27$.

Step 4. Calculate T: $T = W/(W/T) = 0.6/0.27 = 2.22$ m.

Step 5. Find spread above the depressed section: $T_s = 2.22 - 0.6 = 1.62$ m.

Step 6. Use Figure 4.19 to determine Q_s: $Q_s = 0.014$ m^3/s.

Step 7. Compare Q_s (0.014) from Step 6 to the assumed value of Q_s (0.020); since they are not close, try another value of Q_s (e.g., 0.023) and repeat until the calculated and assumed values are close.

The use of Figure 4.21 is illustrated with an example on the figure.
Figure 4.19 can be used to solve the flow for a V-gutter section (utilizing Figure 4.22), such as those in triangular channel sections adjacent to concrete median barriers, using the following steps:

1. Determine S, S_x, n, and Q.
2. Calculate $S_x = (S_{x1} + S_{x2})/(S_{x1} + S_{x2})$.
3. Solve for Q (flow) using Figure 4.19.

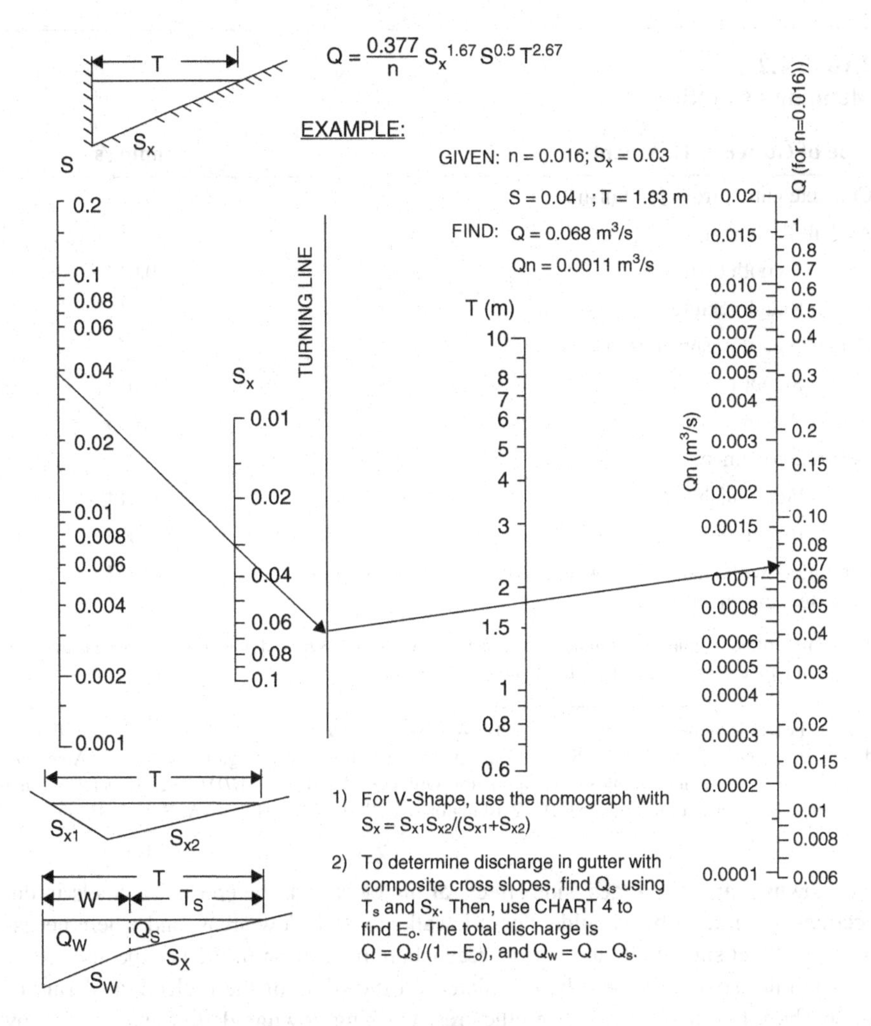

FIGURE 4.19 Nomograph for designing gutter section. (From American Association of State Highway and Transportation Officials, *MDM-SI-2, Model Drainage Manual*, 2000 Metric Edition, AASHTO, Washington, DC, © 2000. Used with permission.)

4.6.1 Inlets

Inlets are provided at regular intervals to collect surface water and convey them to the storm drains. The inlets could be of the grate or curb-opening type or a combination of both. Grate inlets are suitable for continuous grades and should be made bike safe where bike traffic is expected. Curb openings are more suitable for sag points, since they can let in large quantities of water, and could be used where grates are hazardous for bikes or pedestrians. Inlets are spaced at regular intervals, as explained later, and also at points such as sag points in the gutter grade, upstream of median breaks, entrance and exit ramps, crosswalks and intersections, immediately upstream

TABLE 4.2
Manning's Coefficient

Type of Gutter or Pavement	Manning's n
Concrete gutter, troweled finish	0.012
Asphalt Pavement:	
Smooth texture	0.013
Rough texture	0.016
Concrete gutter-asphalt pavement	
Smooth	0.013
Rough	0.015
Concrete pavement	
Float finish	0.014
Broom finish	0.016
For gutters with small slope, where sediment may accumulate, increase above n values by:	0.002

Note: This criteria applies to shoulder widths of 1.8 m or greater. Where shoulder widths are less than 1.8 m, a minimum design spread of 1.8 m should be considered.

Source: Federal Highway Administration (FHWA), *Design Charts for Open-Channel Flow*, Hydraulic design series No. 3 (HDS 3), U.S. Government Printing Office, Washington, DC, 1961; American Association of State and Highway Transportation Officials (AASHTO), *MDM-SI-2, Model Drainage Manual*, 2000, Metric Edition, AASHTO, Washington, DC, 2000. USDOT, FHWA, HDS-3 (1961).

and downstream of bridges, side streets at intersections, the end of channels in cut sections, behind curbs, shoulders or sidewalks to drain low areas, and where necessary to collect snowmelt, and *not* in areas where pedestrian traffic is expected.

The inlet spacing should be calculated on the basis of the collection of runoff. Inlets should be first located from the crest working downgrade to the sag point, by first calculating the distance of the first inlet from the crest and then computing the distances of the other successive inlets. The distance of the first inlet from the crest is calculated as follows:

$$L = \frac{10{,}000 Q_t}{0.0028 CIW} \tag{4.2}$$

Where,

L = the distance of the first inlet from the crest, m
Q_t = the maximum allowable flow, m^3/s
C = the composite runoff coefficient for contributing drainage area
W = the width of contributing drainage area, m
I = the rainfall intensity for design frequency, mm/h

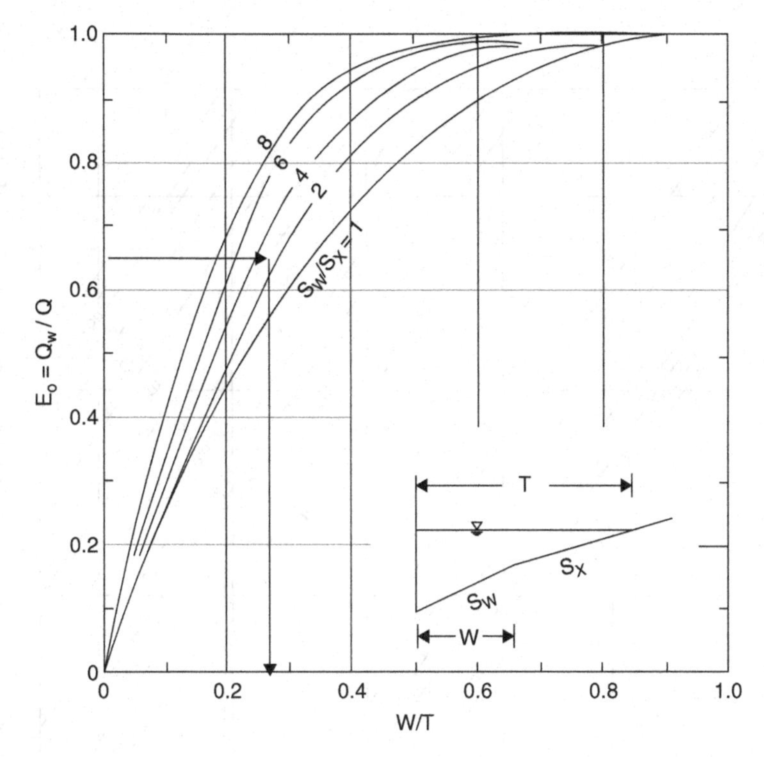

FIGURE 4.20 Ratio of frontal flow to gutter flow. (From American Association of State Highway and Transportation Officials, *MDM-SI-2, Model Drainage Manual*, 2000 Metric Edition, AASHTO, Washington, DC, © 2000. Used with permission.)

4.7 SUBSURFACE DRAINAGE – PERMEABLE BASE PAVEMENTS STRUCTURES

In general, the subsurface drainage system consists of a drainage layer in the subsurface part of the pavement and side drainage channels – the drainage layer of high-permeability material slopes away on both sides to intercept subsurface water and direct it sideways to drainage channels. It is important to provide filters in the drainage layer and the channels such that they remove water and water only, and do not let finer soil particles wash out with the water. The various components are shown in Figure 4.23.

For rigid pavements, there is a separator layer between the subgrade (to prevent migration of fines) and the permeable base over which the PCC layer is placed. In asphalt pavements, the permeable base layer is placed under adequate thickness of the asphalt mix layer.

The resultant slope of the permeable base is given by the following:

$$S_R = (S^2 + S_x^2)^{0.5} \tag{4.3}$$

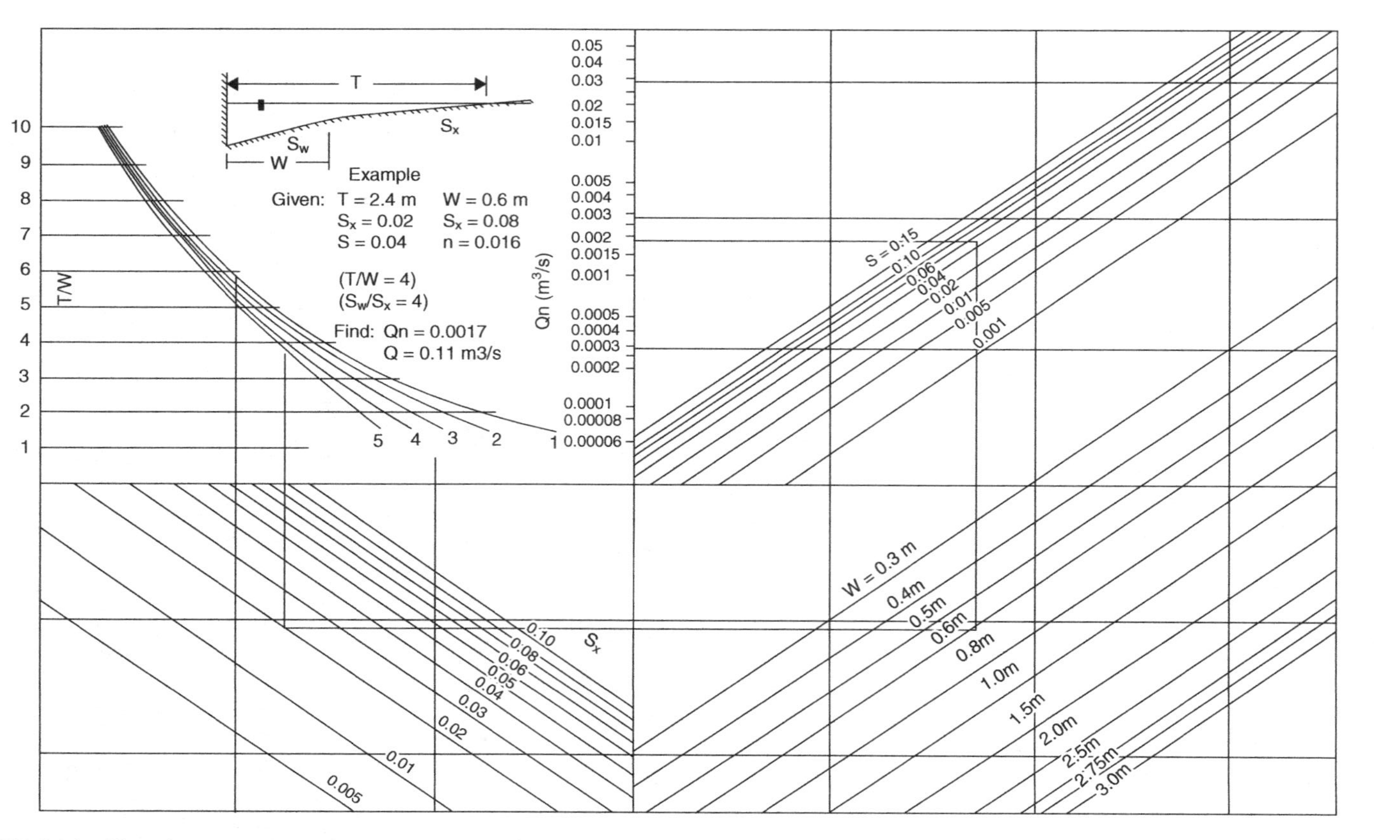

FIGURE 4.21 Flow in composite gutter section. (From American Association of State Highway and Transportation Officials, *MDM-SI-2, Model Drainage Manual*, 2000 Metric Edition, AASHTO, Washington, DC, © 2000. Used with permission.)

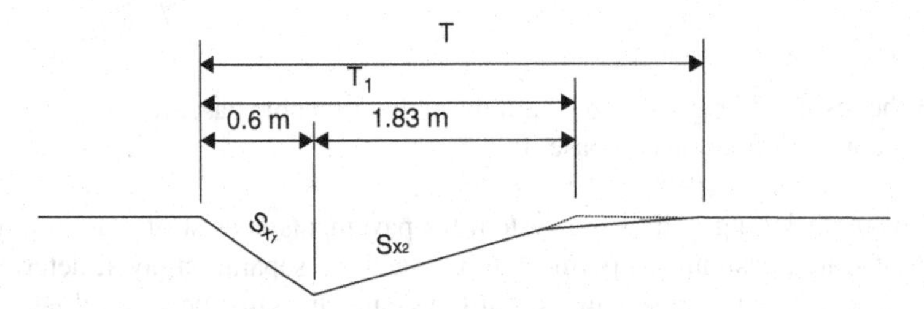

FIGURE 4.22 V-type gutter. (From American Association of State Highway and Transportation Officials, *MDMSI-2, Model Drainage Manual*, 2000 Metric Edition, AASHTO, Washington, DC, © 2000. Used with permission.)

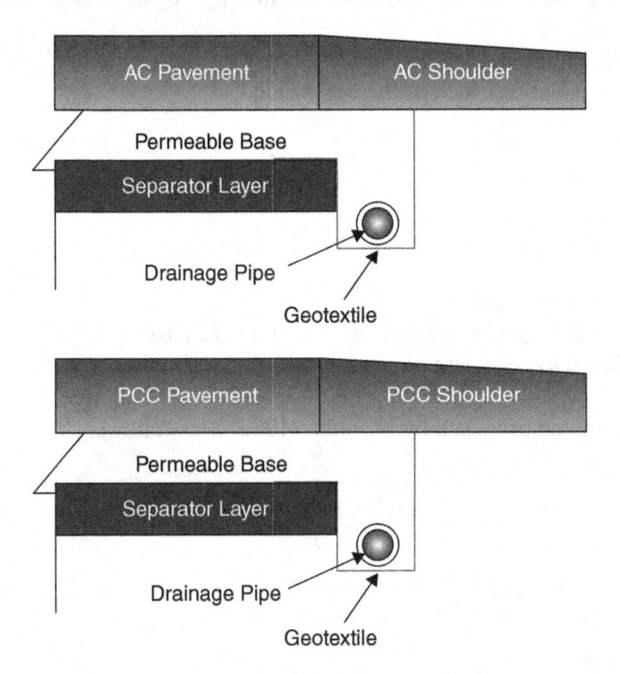

FIGURE 4.23 Subsurface drainage systems in asphalt and concrete pavements.

Where,

S_R = the resultant slope, ft/ft
S = the longitudinal slope, ft/ft
S_x = the cross-slope, ft/ft

The resultant length of the flow path is the following:

$$L_R = W[1+(\frac{S}{S_x})^2]^{0.5} \quad (4.4)$$

Where,

L_R = the resultant length of flow path through permeable base, ft
W = the width of permeable base, ft

The steps in designing a drainage system for pavements consist of determining the amount of water, designing a permeable base and the separation layer, determining the flow to edgedrains and spacing of outflows, and checking the outlet flow.

The coefficient of permeability, k, is an important factor and is primarily dictated by effective grain size, D_{10}, porosity, n, and percentage passing the 0.075 mm sieve. The addition of stabilizer such as asphalt or Portland cement can counterbalance the effect of replacing the fine aggregate portion and hence the loss in stability.

Although k can be determined in the laboratory, quite commonly k is estimated from empirical equations such as the following:

$$k = C_k D_{10}^2 \quad (4.5)$$

Where,

k = the permeability, cm/s
D_{10} = the effective grain size corresponding to size passing 10%
C_k = the Hazen's coefficient, 0.8–1.2

Another equation proposed by Moulton (1980) is as follows:

$$k = \frac{(6.214 * 10^5) D_{10}^{1.478} n^{6.654}}{P_{200}^{0.597}} \quad (4.6)$$

Where,

n = the porosity
P_{200} = the percentage passing the No. 200 sieve

The void ratio, or porosity, has a significant effect on k and the amount of water that can stay within the soil, and this is important since all of the water within the soil cannot be removed, since some of the water would remain as thin film. The porosity that is effective in determining how much water can be removed is called the *effective porosity*.

The total porosity, or porosity, is defined as follows:

$$n = \frac{V_v}{V_T} \quad (4.7)$$

Where,

V_T = the total volume
V_v = the volume of voids
$V_v = V_T - V_S$
V_S = the volume of solid

$$V_v = V_T - \frac{\gamma_d V_T}{\gamma_w G_S} \tag{4.8}$$

G_S = the specific gravity of soil
γ_d = the dry unit weight of soil
γ_w = the unit weight of water

Effective porosity:

$$n_e = \frac{V_v - V_R}{V_T} = n - \frac{V_R}{V_T} \tag{4.9}$$

Where, V_R is the volume of the water retained in the soil.
The volume of water retained in a soil:

$$V_R = \frac{\gamma_d w_c}{\gamma_w} \tag{4.10}$$

Where, w_c is the water content of the soil after draining.
If $V_T = 1$,

$$n_e = n - \frac{\gamma_d w_c}{\gamma_w} \tag{4.11}$$

If a test is conducted to determine the volume of water draining under gravity from a known volume of material, then the effective porosity can be computed as follows:

$$n_e = \frac{V_e}{V_T} \tag{4.12}$$

where V_e is the volume of water draining under gravity.
If W_L is the *water loss* percentage – that is, water drained from the sample – then the effective porosity can be expressed as follows (FHWA, 1992):

$$n_e = \frac{nW_L}{100} \tag{4.13}$$

4.7.1 Design of Permeable Base

The permeable base can be designed according to one of the two available methods. The Moulton (1979) method is based on the idea that the thickness of the base should be equal to or greater than the depth of the flow, which means that the steady flow capacity of the base should be equal to or greater than the rate of inflow. The design equations are as follows:

k = the permeability
S = the slope
L_R = the length of drainage
q_i = the rate of uniform inflow
H_1 = the depth of water at the upper end of the flow path

$$Case\ 1: (S^2 - \frac{4q_i}{k}) < 0$$

$$H_1 = \sqrt{\frac{q_i}{k}} L_R [(\frac{S}{\sqrt{\frac{4q_i}{k} - S^2}})(\tan^{-1} \frac{S}{\sqrt{\frac{4q_i}{k} - S^2}} - \frac{\pi}{2})] \quad (4.14a)$$

$$Case 2: (S^2 - \frac{4q_i}{k}) = 0$$

$$H_1 = \sqrt{\frac{q_i}{k}} L_R^{-1} \quad (4.14b)$$

$$Case 3: (S^2 - \frac{4q_i}{k}) > 0$$

$$H_1 = \sqrt{\frac{q_i}{k}} L_R [\frac{S - \sqrt{S^2 - \frac{4q_i}{k}}}{S + \sqrt{S^2 - \frac{4q_i}{k}}}]^{\frac{S}{2\sqrt{S^2 - \frac{4q_i}{k}}}} \quad (4.14c)$$

The equations can be solved with the use of a chart (Figure 4.24).

The chart can be used to determine the maximum depth of flow, or the required k of the material, when the other parameters are known. The use of the chart is based on the assumption that H_1 equals H_{max}, which is the maximum depth of the flow.

The second approach is based on the concept of time to drain, specifically 50% drainage in ten days, developed on the basis of a study with freeze–thaw-susceptible base courses by Casagrande and Shannon (1952).

For the following equations:

U = the percentage drainage (expressed as a fraction, e.g., 1% = 0.01)
S_1 = the slope factor = H/(LS)

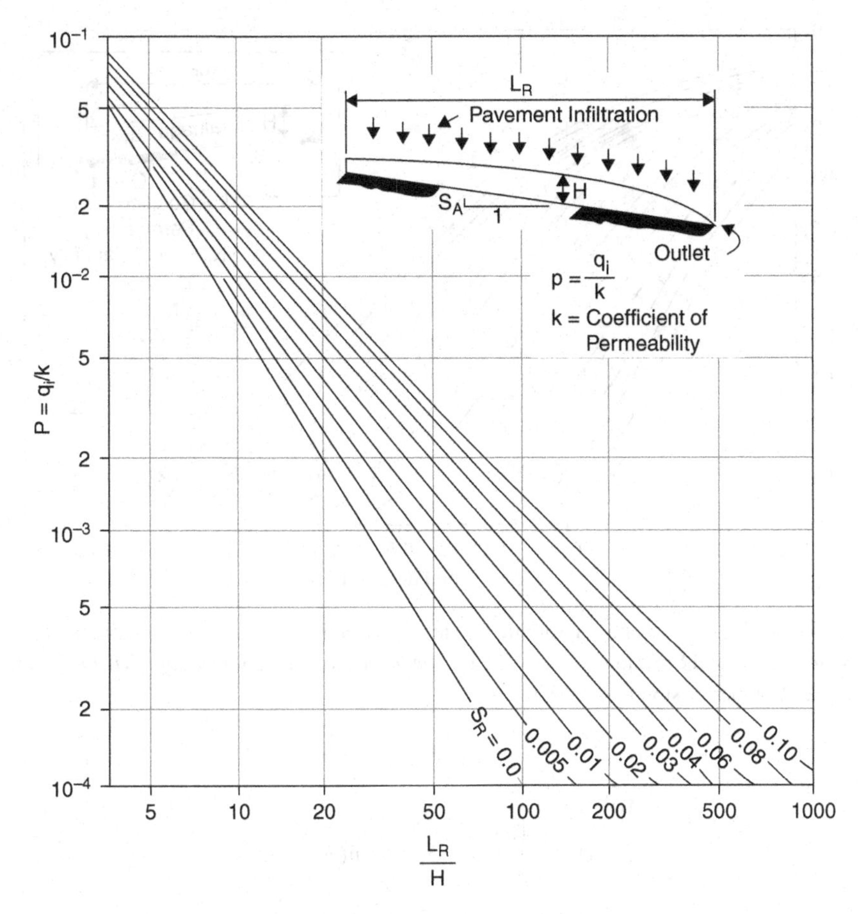

FIGURE 4.24 Nomograph for solving Moulton's (1980) equation. (From Moulton, L.K., Highway subsurface design, FHWA-TS-80–224, U.S. Department of Transportation, Washington, DC, 1980.)

H = the thickness of granular layer
L = the width of granular layer being drained
S = the slope of granular layer
T = the time factor = tkH/n_eL^2
t = the time for drainage, U, to be reached
k = the permeability of granular layer
n_e = the effective porosity of granular material

If U > 0.5,

$$T=(1.2-\frac{0.4}{S_1^{1/3}})[S_1-S_1^2 ln(\frac{S_1+1}{S_1})+S_1 ln(\frac{2S_1-2US_1+1}{(2-2U)(S_1+1)})] \tag{4.15a}$$

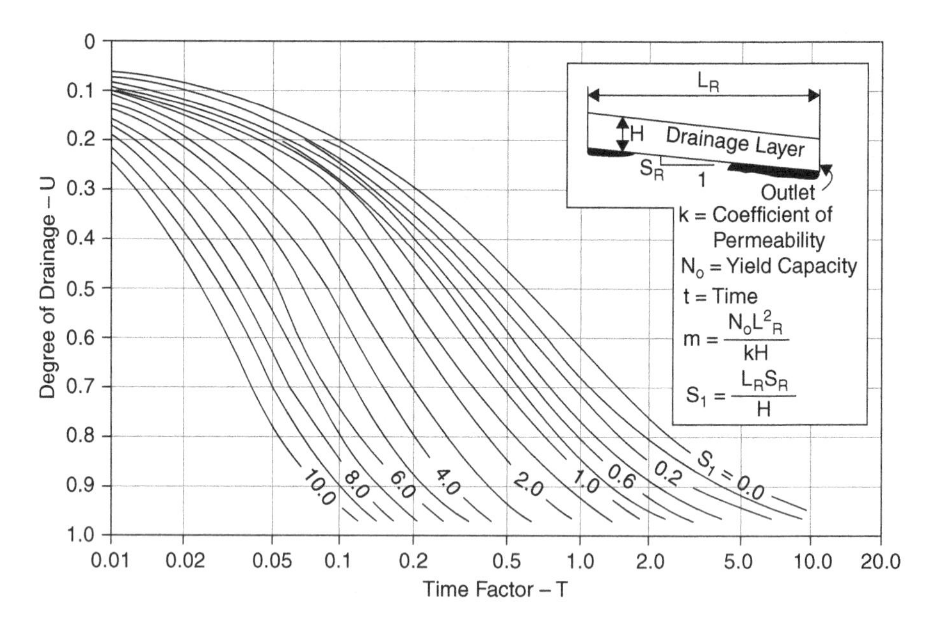

FIGURE 4.25 Nomograph for solving Casagrande and Shannon's (1952) equations. (From Casagrande, A. and Shannon, W.L. 1952. *Proceedings. American Society of Civil Engineers*, 77: 792. With permission from ASCE.)

If $U \leq 0.5$,

$$T = (1.2 - \frac{0.4}{S_1^{1/3}})[2US_1 - S_1^2 \ln(\frac{S_1 + 2U}{S_1})] \tag{4.15b}$$

The equations can be solved by charts as shown in Figure 4.25.

Barber and Sawyer (1952) equations are as follows:

U = the percentage drainage (expressed as a fraction, e.g., 1% = 0.01)
S_1 = the slope factor = H/DS
H = the thickness of granular layer
D = the width of granular layer being drained
S = the slope of granular layer
T = the time factor = $(tkH)/(n_e L^2)$
t = the time for drainage, U, to be reached
k = the permeability of granular layer
n_e = the effective porosity of granular material

For $0.5 \leq U \leq 1.0$,

$$T = 0.5S_1 - 0.48S_1^2 \log(1 + \frac{2.4}{S_1}) + 1.15S_1 \log[\frac{S_1 - US_1 + 1.2}{(1-U)(S_1 + 2.4)}] \tag{4.16a}$$

TABLE 4.3
Unstabilized Permeable Base Gradations

State	Percent Passing Sieve Size											
	2 in	1 ½ in	1 in	¾ in	½ in	3/8 in	No. 4	No. 8	No. 16	No. 40	No. 50	No. 200
AASHTO #57		100	95–100		25–60		0–10	0–5				
AAHSTO #67			100	90–100		20–55	0–10	0–5				
Iowa			100					10–35			0–15	0–6
Minnesota			100	65–100		35–70	20–45			2–10		0–3
New Jersey		100	95–100		60–80		40–55	5–25	0–12		0–5	
Pennsylvania	100			52–100		33–65	8–40		0–12			0–5

Source: Federal Highway Administration (FHWA), Pavement subsurface drainage design, FHWA-HI-99-028, NHI Course No. 131026, U.S. Department of Transportation, Washington, DC, 1999.

For $0 \leq U \leq 0.5$,

$$T = US - 0.48 S_1^2 \log\left(1 + \frac{4.8U}{S_1}\right) \tag{4.16b}$$

4.7.1.1 Materials for Permeable Base

The time to drain of a soil depends primarily on its coefficient of permeability, which is primarily dependent on the aggregate gradation. Gradations of typical materials (as suggested by FHWA, 1999) are given in Tables 4.3–4.5. Generally, materials with permeability exceeding 1,000 ft/day are used for highways.

4.7.2 Design of Separator or Filter Layer

The separator layer can be made up of aggregates or a geotextile layer. The functions of this layer include preventing pumping of fines from the subgrade to the permeable base, providing a stable platform to facilitate the construction of the overlying layers, directing water infiltrating from above to the side drains or edgedrains and preventing it from entering the subgrade, and distributing the loads over the subgrade without deflecting excessively. Separator layers with aggregates range in thickness from 4–12 inches and can provide the stable construction platform as well as distribute loads over the subgrade without deflecting. Generally, geotextiles are used over stabilized subgrades. The separator layer should be such that subgrade fines do not move up to the separator layer and the fines from the separator layer do not move into the permeable base. The following requirements must be met:

D_{15} (separator layer) ≤ 5 D_{85} (subgrade)
D_{50} (separator layer) ≤ 25 D_{50} (subgrade)
D_{15} (base) ≤ 5 D_{85} (separator layer)

TABLE 4.4
Typical Asphalt Stabilized Permeable Base Gradations

State	Percent Passing Sieve Size					
	1 in	½ in	3/8 in	No. 4	No. 8	No. 200
California	100	90–100	20–45	0–10		0–2
Florida	100	90–100	20–45	0–10	0–5	0–2
Illinois	90–100	84–100	40–60	0–12		
Kansas	100	90–100	20–45	0–10	0–5	0–2
Ohio	95–100			25–60	0–10	
Texas	100	95–100	20–45	0–15	0–5	2–4
Wisconsin	95–100	80–95	25–50	35–60	20–45	3–10
Wyoming	90–100	20–50		20–50	10–30	0–4

Source: Federal Highway Administration (FHWA), Pavement subsurface drainage design, FHWA-HI-99-028, NHI Course No. 131026, U.S. Department of Transportation, Washington, DC, 1999.

TABLE 4.5
Typical Cement-Stabilized Permeable Base Gradations

State	Percent Passing Sieve Size						
	1 1/2 in.	1 in.	3/4 in.	1/2 in.	3/8 in.	No. 4	No. 8
California	100	88–100	X ± 15		X ± 15	0–16	0–6
Virginia		100		25–60		0–10	0–5
Wisconsin		100	90–100		20–55	0–10	0–5

Note: X is the percentage submitted by the contractor.

Source: Federal Highway Administration (FHWA), Pavement subsurface drainage design, FHWA-HI-99-028, NHI Course No. 131026, U.S. Department of Transportation, Washington, DC, 1999.

D_{50} (base) ≤25 D_{50} (separator layer)

The requirements of the aggregate separator layer are as follows:

1. Should have durable, crushed, angular aggregate.
2. Maximum Los Angeles abrasion loss of 50%.
3. Maximum soundness of 12% or 18% loss as determined by the sodium sulfate or magnesium sulfate tests.
4. Density should be at least 95% of the maximum density.
5. Maximum percentage of materials passing the No. 200 sieve is 12%.
6. Coefficient of uniformity is between 20 and 40.

TABLE 4.6
Typical Gradation requirements for Separator Layer

Sieve Size	Percent Passing
$1^1/_2$ in	100
$^3/_4$ in	95–100
No. 4	50–80
No. 40	20–35
No. 200	5–12

Source: Federal Highway Administration (FHWA), Drainage requirements in pavements (DRIP), Developed by Applied Research Associates for the Federal Highway Administration, Contract No. DTFH61-00-F-00199, U.S. Department of Transportation, Washington, DC, 2001.

Examples of typical gradations used by state highway agencies are provided in Table 4.6.

4.7.2.1 Pipes

Water through a trench or under a fill is conveyed through a drainage pipe. There are different materials used for drainage pipes, as discussed below.

Concrete pipes are used in the high acidic or alkaline conditions. Type II or Type V cements can be used where water/soil is with a high concentration of sulfate. In the event where the culvert is subjected to tidal damage or saltwater spray, high density concrete pipes are recommended. Plastic lined concrete pipes are suitable for carrying highly corrosive substances.

Corrugated steel pipes are used where solvents are expected to enter the pipe. These pipes are galvanized and coated with asphalt. Asphalt coating is normally provided where the steel pipe is subjected to continuous wetness or flow of water.

Rigid or plastic pipes are used where there are harmful substances such as spilled petroleum products or industrial wastes, which can be harmful to the asphalt and coating in corrugated metal pipes. Plastic pipes can be used for low-volume road applications. In the case airfield pavements, these pipes should comply with Item D-701, Pipe for storm Drains and Culverts in AC 150/5370-10 standards for construction of airports.

Corrugated aluminum alloy pipes are used where sea water is present due to its corrosion resistive properties. These pipes are not used where the soils are highly acidic ($pH < 5$) or alkaline ($pH > 9$). Also, these are not used in material classified as OH (organic clays/silts of medium to high plasticity) or OL (organic silts/clays of low plasticity) according to the Unified Soil Classification System (ASTM D2487).

Various considerations are taken during the pipe selection which includes best slope, proper strength, full capacity, least depth and provision of material installation, so that each system is designed economically. Also proper bedding, backfilling, compaction and preventing backfill material from infiltrating into pipe are very important. Different factors to be considered in the selection of the type of pipe and its installation include the following:

Strength: The pipe should have ample strength under loads. Tables 4.7–4.14 show suggested minimum and maximum cover for various pipe installations. Note that here the backfill material is compacted to at least 90% of ASTM D 1557 or AASHTO T99 density. The guidelines for minimum cover to protect the pipe at the time of construction are shown in Table 4.15.

Bedding of pipe: Figures 4.26–4.29 show the bedding for different pipes. Details of different classes of conduit installation which influence loads of underground conduits are provided in FAA, 2013.

Consideration of frost conditions: Heaving of frost-susceptible soils under drains and culverts causes displacement of pipes with consequent loss of alignment, joint failures and even pipe breakage. The detrimental effects of heaving of frost-susceptible soils around and under drains and culverts are considered in the design of drainage systems in cold areas.

The placement of drains and culverts beneath pavements should be avoided and in cases, when such placement is unavoidable, the pipes should be installed before the laying the base course. Excavation through base courses and laying pipes can cause non-uniformity between the compacted trench backfill and the adjacent material.

In the case of frost-susceptible subgrades where the highest groundwater table is 5 feet or more below the maximum depth of frost penetration and the pipe diameter is 18 inch or more: i) the centerline of the pipe should be placed at or below the depth of maximum frost penetration and backfill material around the pipe should be highly free-draining and non-frost-susceptible material; or ii) place the centerline of the pipe one-third diameter below the depth of maximum frost penetration. The invert of the pipe should be placed at or below the depth of maximum frost penetration in order to prevent water from freezing in the pipe. Water freezing in culverts is a serious issue in arctic and subarctic regions. The number of drainage structures should be held to a minimum and should be designed based on twice the normal design capacity. Thawing devices can be provided in all culverts up to 48 inches in diameter. Large-diameter culverts have to be cleaned manually immediately prior to the spring thaw.

Soil infiltration through pipes: Infiltration of fine-grained soils and flowing of dispersive soils causes a serious threat to the drainage facilities. In order to prevent infiltration, watertight jointing (watertight flexible joint materials in rigid pipe and with watertight coupling bands in flexible pipe) is necessary in culverts and drains placed on steep slopes. Culverts and drains placed on steep slopes should be large enough and properly vented in order to prevent full pipe flow. This helps in maintaining the hydraulic gradient above the pipe invert but below crown of the pipe, thereby reducing the infiltration of soil water through joints. In the case of pavements constructed on fine-grained subgrades and subject to perched water table conditions, pervious

TABLE 4.7
Suggested Maximum Cover Requirements for Concrete Pipe, Reinforced Concrete, H-20 Highway Loading

Diameter (inch)	Suggested Maximum Cover Above Top of Pipe, feet Circular Section Class				
	1500	2000	2500	3000	3750
12	9	13	16	19	24
24	10	13	17	19	24
36	10	13	17	20	25
48	10	13	17	20	25
60	10	14	17	20	25
72	10	14	17	20	25
84	11	14	17	21	24
108	11	14	17	21	26

Non-reinforced Concrete

Diameter (inch)	Suggested Maximum Cover Above Top of Pipe, feet Circular Section		
	I	II	III
12	14	14	17
24	13	13	14
36	9	12	12

Notes:
1. The suggested values shown are for average conditions and are to be considered as guidelines only for dead load plus H-20 live load.
2. Soil conditions, trench width, and bedding conditions vary widely throughout varying climatic and geographical areas.
3. Calculations to determine maximum cover should be made for all individual pipe and culvert installations underlying roads, streets, and open storage areas subject to H-20 live loads. Cooper E-80 railway loadings should be independently made.
4. Cover depths are measured from the bottom of the subbase of pavements, or the top of unsurfaced areas, to the top of the pipe.
5. Calculations to determine maximum cover for Cooper E-80 railway loadings are measured from the bottom of the tie to the top of the pipe.
6. "D" loads listed for the various classes of reinforced-concrete pipe are the minimum required 3-edge test loads to produce ultimate failure in pounds per linear foot of interval pipe diameter.
7. Each diameter pipe in each class designation of non-reinforced concrete has a different D-load value that increases with wall thickness.
8. If apipe produced by a manufacturer exceeds the strength requirements established by indicated standards, cover depths may be adjusted accordingly.
9. See Table 4.15 for suggested minimum cover requirements.

Source: FAA, 2013.

TABLE 4.8
Suggested Maximum Cover Requirements for Corrugated Aluminum Alloy Pipe, Riveted, Helical, or Welded Fabrication 2.66-inch Spacing, 0.5-inch-deep Corrugations, H-20 Highway Loading

Diameter (inch)	Suggested Maximum Cover Above Top of Pipe, feet									
	Circular Section Thickness (inch)					Vertically Elongated Section Thickness (inch)				
	0.060	0.075	0.105	0.135	0.164	0.060	0.075	0.105	0.135	0.164
12	50	50	86	90	93					
15	40	40	69	72	74					
18	33	33	57	60	62					
24	25	25	43	45	46					
30	20	20	34	36	37					
36	16	16	28	30	31					
42	16	16	28	30	31			50	52	53
48			28	30	31			43	45	47
54			28	30	31					
60				30	31					
66					31					
72					31					

Notes:

1. Corrugated aluminum alloy pipe will conform to the requirements of ASTM B745/B745M.
2. The suggested values shown are for average conditions and are guidelines only for dead load plus H-20 live load. Cooper E-80 railway loadings should be independently made.
3. Soil conditions, trench width and bedding conditions vary widely throughout varying climatic and geographical areas.
4. Calculations to determine maximum cover should be made for all individual pipe and culvert installations underlying roads, streets and open storage areas subject to H-20 live loads.
5. Cover depths are measured from the bottom of the subbase of pavements, or the top of unsurfaced areas, to the top of the pipe.
6. Calculations to determine maximum cover for Cooper E-80 railway loadings are measured from the bottom of the tie to the top of the pipe.
7. Vertical elongation will be accomplished by shop fabrication and will usually be 5% of the pipe diameter.
8. See Table 4.15 for suggested minimum cover requirements.

Source: FAA, 2013.

TABLE 4.9
Suggested Maximum Cover Requirements for Corrugated Steel Pipe, 2.66-inch Spacing, 0.5-inch-deep Corrugations

Diameter (inch)	H-20 Highway Loading					
	Suggested Maximum Cover Above Top of Pipe, feet					
	Helical – Thickness (inch)					
	0.052	0.064	0.079	0.109	0.138	0.168
12	170	213	266	372		
15	136	170	212	298		
18	113	142	173	212		
21	97	121	139	164		
24	85	106	120	137	155	
27	75	94	109	120	133	
30	68	85	101	110	119	
36	56	71	88	98	103	
42	48	60	76	92	95	99
48		53	66	88	91	93
54			59	82	88	90
60				74	86	87
66					85	86
72					79	85
78						84
84						75

Notes:
1. Corrugated steel pipe will conform to the requirements of ASTM A760/A760M, ASTM A761/A761M, ASTM A762/A762M, and ASTM A849.
2. The suggested maximum heights of cover shown in the tables are calculated on the basis of the current AASHTO Standard Specifications for Highway Bridges and are based on circular pipe.
3. Soil conditions, trench width and bedding conditions vary widely throughout varying climatic and geographical areas.
4. Calculations to determine maximum cover should be made for all individual pipe and culvert installations underlying roads, streets, and open storage areas subject to H-20 live loads. Cooper E-80 railway loadings should be independently made.
5. Cover depths are measured from the bottom of the subbase of pavements, or the top of unsurfaced areas, to the top of the pipe.
6. Calculations to determine maximum cover for Cooper E-80 railway loadings are measured from the bottom of the tie to the top of the pipe.
7. If a pipe produced by a manufacturer exceeds the strength requirements established by indicated standards, then cover depths may be adjusted accordingly.
8. See Table 4.15 for suggested minimum cover requirements.

Source: FAA, 2013.

TABLE 4.10
Suggested Maximum Cover Requirements for Structural Plate Aluminum Alloy Pipe, 9-inch Spacing, 2.5-inch Corrugations

Diameter (inch)	H-20 Highway Loading Suggested Maximum Cover Above Top of Pipe, feet Circular Section						
	Thickness (inch)						
	0.10	0.125	0.15	0.175	0.20	0.225	0.250
72	24	32	41	48	55	61	64
84	20	27	35	41	47	52	55
96	18	24	30	36	41	45	50
108	16	21	27	32	37	40	44
120	14	19	24	29	33	36	40
132	13	17	22	26	30	33	36
144	12	16	20	24	27	30	33
156		14	18	22	25	28	30
168		13	17	20	23	26	28
180			16	19	22	24	26

Notes:
1. Structural plate aluminum alloy pipe will conform to the requirements of ASTM B745/B745M.
2. Soil conditions, trench width and bedding conditions vary widely throughout varying climatic and geographical areas.
3. Calculations to determine maximum cover should be made for all individual pipe and culvert installations underlying roads, streets and open storage areas subject to H-20 live loads. Cooper E-80 railway loadings should be independently made.
4. Cover depths are measured from the bottom of the subbase of pavements, or the top of unsurfaced areas, to the top of the pipe.
5. Calculations to determine maximum cover for Cooper E-80 railway loadings are measured from the bottom of the tie to the top of the pipe.
6. If a pipe produced by a manufacturer exceeds the strength requirements established by indicated standards, cover depths may be adjusted accordingly.
7. See Table 4.15 for suggested minimum cover requirements.

Source: FAA, 2013.

base courses with a minimum thickness of about 6 inch with provisions for drainage should be provided.

4.7.2.2 Geotextile Separator Layer

The requirements for geotextile separators have been set as follows (FHWA, 1998):

$$AOS\ or\ O_{95\ (geotextile)} \leq B\,D_{85\ (soil)} \tag{4.17}$$

TABLE 4.11
Suggested Maximum Cover Requirements for Corrugated Steel Pipe, 5-inch Span, 1-inch-deep Corrugations

Diameter (inch)	H-20 Highway Loading Suggested Maximum Cover Above Top of Pipe, feet Helical –Thickness (inch)				
	0.064	0.079	0.109	0.138	0.168
48	54	68	95	122	132
54	48	60	84	109	117
60	43	54	76	98	107
66	39	49	69	89	101
72	36	45	63	81	96
78	33	41	58	75	92
84	31	38	54	70	85
90	29	36	50	65	80
96		34	47	61	75
102		32	44	57	70
108			42	54	66
114			40	51	63
120			38	49	60

Notes:
1. Corrugated steel pipe will conform to the requirements of ASTM A760/A760M, ASTM A761/A761M, ASTM A762/A762M, and ASTM A849.
2. The suggested maximum heights of cover shown in the table are calculated on the basis of the current AASHTO Standard Specifications for Highway Bridges and are based on circular pipe.
3. Soil conditions, trench width and bedding conditions vary widely throughout varying climatic and geographical areas.
4. Calculations to determine maximum cover should be made for all individual pipe and culvert installations underlying roads, streets, and open storage areas subject to H-20 live loads. Cooper E-80 railway loadings should be independently made.
5. Cover depths are measured from the bottom of the subbase of pavements, or the top of unsurfaced areas, to the top of the pipe.
6. Calculations to determine maximum cover for Cooper E-80 railway loadings are measured from the bottom of the tie to the top of the pipe.
7. If pipe produced by a manufacturer exceeds the strength requirements established by indicated standards, cover depths may be adjusted accordingly.
8. See Table 4.15 for suggested minimum cover requirements.

Source: FAA, 2013.

TABLE 4.12
Suggested Maximum Cover Requirements for Structural Plate Steel Pipe, 6-inch Span, 2-inch-deep Corrugations

Diameter (ft)	H-20 Highway Loading						
	Suggested Maximum Cover Above Top of Pipe, feet Thickness (inch)						
	0.109	0.138	0.168	0.188	0.218	0.249	0.280
5.0	46	68	90	103	124	146	160
5.5	42	62	81	93	113	133	145
6.0	38	57	75	86	103	122	133
6.5	35	52	69	79	95	112	123
7.0	33	49	64	73	88	104	114
7.5	31	45	60	68	82	97	106
8.0	29	43	56	64	77	91	100
8.5	27	40	52	60	73	86	94
9.0	25	38	50	57	69	81	88
9.5	24	36	47	54	65	77	84
10.0	23	34	45	51	62	73	80
10.5	22	32	42	49	59	69	76
11.0	21	31	40	46	56	66	72
11.5	20	29	39	44	54	63	69
12.0	19	28	37	43	51	61	66
12.5	18	27	36	41	49	58	64
13.0	17	26	34	39	47	56	61
13.5	17	25	33	38	46	54	59
14.0	16	24	32	36	44	52	57
14.5	16	23	31	35	42	50	55
15.0	15	22	30	34	41	48	53
15.5	15	22	29	33	40	47	51
16.0		21	28	32	38	45	50
16.5		20	27	31	37	44	48
17.0		20	26	30	36	43	47
17.5		19	25	29	35	41	45
18.0			25	28	34	40	44
18.5			24	27	33	39	43
19.0			23	27	32	38	42
19.5			23	26	31	37	41
20.0				25	31	36	40

TABLE 4.12
(Cont.)

Diameter (ft)	H-20 Highway Loading						
	Suggested Maximum Cover Above Top of Pipe, feet Thickness (inch)						
	0.109	0.138	0.168	0.188	0.218	0.249	0.280
20.5				25	30	35	39
21.0					29	34	38
21.5					28	34	37
22.0					28	33	36
22.5					27	32	35
23.0						31	34
23.5						31	34
24.0						30	33
24.5							32
25.0							32
25.5							31

Notes:

1. Corrugated steel pipe will conform to the requirements of ASTM A760/A760M, ASTM A761/A761M, ASTM A762/A762M, and ASTM A849.
2. The suggested maximum heights of cover shown in the table are calculated on the basis of the current AASHTO Standard Specifications for Highway Bridges and are based on circular pipe.
3. Soil conditions, trench width and bedding conditions vary widely throughout varying climatic and geographical areas.
4. Calculations to determine maximum cover should be made for all individual pipe and culvert installations underlying roads, streets, and open storage areas subject to H-20 live loads. Cooper E-80 railway loadings should be independently made.
5. Cover depths are measured from the bottom of the subbase of pavements, or the top of unsurfaced areas, to the top of the pipe.
6. Calculations to determine maximum cover for Cooper E-80 railway loadings are measured from the bottom of the tie to the top of the pipe.
7. If pipe produced by a manufacturer exceeds the strength requirements established by indicated standards, cover depths may be adjusted accordingly.
8. See Table 4.15 for suggested minimum cover requirements.

Source: FAA, 2013.

Where,

AOS = the apparent opening size, mm
O_{95} = the opening size in the geotextile for which 95% are smaller, mm
$AOS \approx O_{95}$
B = dimensionless coefficient
D_{85} = the soil particle size for which 85% are smaller, mm

TABLE 4.13
Suggested Maximum Cover Requirements for Corrugated Steel Pipe, 3-inch Span, 1-inch Corrugations

Diameter (inch)	H-20 Highway Loading — Suggested Maximum Cover Above Top of Pipe, feet									
	Riveted – Thickness (inch)					Helical – Thickness (inch)				
	0.064	0.079	0.109	0.138	0.168	0.064	0.079	0.109	0.138	0.168
36	53	66	98	117	130	81	101	142	178	201
42	45	56	84	101	112	69	87	122	142	157
48	39	49	73	88	98	61	76	107	122	132
54	35	44	65	78	87	54	67	95	110	117
60	31	39	58	70	78	48	61	85	102	107
66	28	36	53	64	71	44	55	77	97	101
72	26	33	49	58	65	40	50	71	92	96
78	24	30	45	54	60	37	47	65	84	93
84	22	28	42	50	56	34	43	61	78	91
90	21	26	39	47	52	32	40	57	73	89
96		24	36	44	49		38	53	69	84
102		23	34	41	46		35	50	64	79
108			32	39	43			47	61	75
114			30	37	41			45	58	71
120			29	35	39			42	55	67

Notes:

1. Corrugated steel pipe will conform to the requirements of ASTM A760/A760M, ASTM A761/A761M, ASTM A762/A762M, and ASTM A849.
2. The suggested maximum heights of cover shown in the table are calculated on the basis of the current AASHTO Standard Specifications for Highway Bridges and are based on circular pipe.
3. Soil conditions, trench width and bedding conditions vary widely throughout varying climatic and geographical areas.
4. Calculations to determine maximum cover should be made for all individual pipe and culvert installations underlying roads, streets, and open storage areas subject to H-20 live loads. Cooper E-80 railway loadings should be independently made.
5. Cover depths are measured from the bottom of the subbase of pavements, or the top of unsurfaced areas, to the top of the pipe.
6. Calculations to determine maximum cover for Cooper E-80 railway loadings are measured from the bottom of the tie to the top of the pipe.
7. If pipe produced by a manufacturer exceeds the strength requirements established by indicated standards, cover depths may be adjusted accordingly.
8. See Table 4.15 for suggested minimum cover requirements.

Source: FAA, 2013.

TABLE 4.14
Suggested Guidelines for Minimum Cover

Pipe	H-20 Highway Loading		
	Minimum Cover to Protect Pipe		Minimum Finished Height of Cover (From Bottom of Subbase to Top of Pipe)
	Pipe Diameter (inch)	Height of Cover During Construction (feet)	
Concrete pipe reinforced	12–108	Diameter/2 or 3.0 ft, whichever is greater	Diameter/2 or 2.0 ft, whichever is greater
Non-reinforced	12–36	Diameter/2 or 3.0 ft, whichever is greater	Diameter/2 or 2.0 ft, whichever is greater
Corrugated aluminum pipe 2.66 inch by 0.5 inch	12–24, 30 and over	1.5 ft Diameter	Diameter/2 or 1.0 ft, whichever is greater Diameter/2
Corrugated steel pipe 3 inch by 1 inch	12–30, 36 and over	1.5 ft Diameter	Diameter/2 or 1.0 ft, whichever is greater Diameter/2
Structural plate aluminum alloy pipe 9 inch by 2.5 inch	72 and over	Diameter/2	Diameter/4
Structural plate steel 6 inch by 2 inch	60 and over	Diameter/2	Diameter/4

Notes:

1. All values shown above are for average conditions and are guidelines only.
2. Calculations should be made for minimum cover for all individual pipe installation for pipe underlying roads, streets and open storage areas subject to H-20 live loads.
3. Calculations for minimum cover for all pipe installations should be made separately for all Cooper E-80 railroad live loading.
4. In seasonal frost areas, minimum pipe cover must meet local code requirements for protection of storm drains.
5. Pipe placed under rigid pavement will have minimum cover from the bottom of the subbase to the top of pipe of 1.0 ft for pipe up to 60 inch and greater than 1.0 ft for sizes above 60 inch if calculations so indicate.
6. Trench widths depend upon varying conditions of construction but may be as wide as is consistent with the space required to install the pipe and as deep as can be managed from practical construction methods.
7. Non-reinforced concrete pipe is available in sizes up to 36 inch.
8. See Table 4.15 for suggested minimum cover requirements.

Source: FAA, 2013.

TABLE 4.15
Minimum Depth of Cover in Feet for Pipe Under Flexible Pavement

CORRUGATED ALUMINUM 2 2/3" × 1/2" or 2" × 1/2" CORRUGATIONS

AIRCRAFT WHEEL LOAD–Up to 30,000 lb. single and up to 40,000 lb. dual

Metal thickness (in.)	Pipe diameter (in.)								
	12	18	24	36	48	60	72	84	96
0.060.........	2.0	2.5	2.5						
0.075.........	1.5	2.0	2.5	2.5	3.0				
0.105.........		1.5	1.5	1.5	2.0	2.5	3.0		
0.135.........			1.0	1.0	1.5	1.5	1.5		
0.165.........					1.0	1.5	1.5	2.0	2.0

AIRCRAFT WHEEL LOAD–40,000 lb. dual to 110,000 lb. dual

Metal thickness (in.)	Pipe diameter (in.)								
	12	18	24	36	48	60	72	84	96
0.060.........	2.0	2.5	2.5						
0.075.........	1.5	2.0	2.5	2.5	3.0				
0.105.........		1.5	1.5	1.5	2.0	2.5	3.0		
0.135.........				1.5	1.5	2.0	2.5	3.0	
0.165.........					1.5	1.5	2.0	2.0	2.5

AIRCRAFT WHEEL LOAD–110,000 lb. dual to 200,000 lb. dual; 190,000 lb. dt. to 350,000 lb. dt.; up to 750,000 lb. ddt. & 1,500,000 lb.

Metal thickness (in.)	Pipe diameter (in.)								
	12	18	24	36	48	60	72	84	96
0.060.........	3.0	3.0	3.0						
0.075.........	3.0	3.0	3.0	3.5	5.0				
0.105.........		2.0	2.0	2.5	3.5	4.5			
0.135.........				2.0	3.0	4.0	4.5	5.5	
0.165.........					2.5	3.5	4.0	5.0	5.5

CORRUGATED ALUMINUM 6" × 1" CORRUGATIONS

AIRCRAFT WHEEL LOAD–Up to 30,000 lb. single and up to 40,000 lb. dual

Metal thickness (in.)	Pipe diameter (in.)							
	36	48	60	72	84	96	108	120
0.060.................	2.0	2.0	2.5	3.0				
0.075.................	1.0	1.5	2.0	2.5	3.5			
0.105.................	1.0	1.0	1.5	2.0	3.0	3.5		
0.135.................			1.5	2.0	2.5	3.0	4.0	
0.165.................					2.0	2.5	3.5	4.5

AIRCRAFT WHEEL LOAD–40,000 lb. dual to 110,000 lb. dual

Metal thickness (in.)	Pipe diameter (in.)							
	36	48	60	72	84	96	108	120
0.060.................	2.5	3.0	3.5	4.0				
0.075.................	1.5	2.0	2.5	3.0	4.0			
0.105.................	1.5	1.5	2.0	2.5	3.5	4.0		
0.135.................			2.0	2.5	3.0	3.5	4.5	
0.165.................					2.5	3.0	4.0	5.0

AIRCRAFT WHEEL LOAD–110,000 lb. d. to 200,000 lb. d; 190,000 lb. dt, to 350,000 lb. dt.; up to 750,000 lb. ddt. & 1,500,000 lb.

Metal thickness (in.)	Pipe diameter (in.)							
	36	48	60	72	84	96	108	120
0.060.................	4.0	4.5	5.0	5.0				
0.075.................	3.0	3.5	3.5	4.0	4.0			
0.105.................	2.0	2.0	3.0	2.5	4.0	4.5		
0.135.................			2.5	3.0	3.5	4.0	5.0	
0.165.................					3.0	3.5	4.5	5.5

CLAY

AIRCRAFT WHEEL LOAD–up to 30,000 lb. single and up to 40,000 lb. dual

Pipe type	Pipe diameter (in.)								
	6	10	12	15	18	21	24	30	36
Std. strength clay.....	2.0	2.5	2.5	2.5	2.5	2.5	2.5	2.5	2.5
Extra strength day	2.0	2.0	2.0	2.0	2.0	2.0	2.0	2.0	2.0

AIRCRAFT WHEEL LOAD–40,000 lb. dual to 110,000 lb. dual

Pipe type	Pipe diameter (in.)								
	6	10	12	15	18	21	24	30	36
Std. strength clay.....	4.0	5.5	6.0	6.0	6.0	6.0	6.0	6.0	6.0
Extra strength day	2.0	3.5	3.5	3.5	3.5	3.5	3.5	3.5	3.5

ASBESTOS CEMENT

AIRCRAFT WHEEL LOAD–up to 30,000 lb. single and up to 40,000 lb. dual

Asbestos cement–class	Pipe diameter (in.)								
	6	10	12	16	18	24	30	36	42
1500..........	2.5	2.5	2.5	2.5					
2400..........	2.5	2.5	2.5	2.5	2.5	2.5			
3300..........	1.5	1.5	1.5	1.5	1.5	1.5	1.5		
4000..........		1.5	1.5	1.5	1.5	1.5	1.5	1.5	
5000..........		1.5	1.5	1.5	1.5	1.5	1.5	1.5	1.5
6000..........								1.0	1.0
7000..........								1.0	1.0

AIRCRAFT WHEEL LOAD–40,000 lb. dual to 110,000 lb. dual

Asbestos cement–class	Pipe diameter (in.)								
	6	10	12	16	18	24	30	36	42
1500..........	5.5	5.5	5.5	5.5					
2400..........	6.0	6.0	6.0	6.0	6.0	6.0			
3300..........	3.5	3.5	3.5	3.5	3.5	3.5			
4000..........		3.5	3.5	3.5	3.5	3.5	3.5	3.5	
5000..........		3.5	3.5	3.5	3.5	3.5	3.5	3.5	3.5
6000..........								2.5	2.5
7000..........								2.5	2.5

TABLE 4.15
(Cont.)

CORRUGATED STEEL 2 2/3" × 1/2" CORRUGATIONS

AIRCRAFT WHEEL LOAD–Up to 30,000 lb. single and up to 40,000 lb. dual

Metal thickness (in.)	Pipe diameter (in.)								
	12	18	24	36	48	60	72	84	96
0.052.........	1.0	1.0	1.5	1.5					
0.064.........	1.0	1.0	1.0	1.5	1.5				
0.079.........	1.0	1.0	1.0	1.5	1.5	1.5			
0.109.........			1.0	1.0	1.0	1.0	1.5		
0.138.........				1.0	1.0	1.0	1.0	1.5	
0.168.........				1.0	1.0	1.0	1.0	1.5	1.5

AIRCRAFT WHEEL LOAD–40,000 lb. dual to 110,000 lb. dual

Metal thickness (in.)	Pipe diameter (in.)								
	12	18	24	36	48	60	72	84	96
0.052.........	1.5	2.0	2.0	2.5					
0.064.........	1.5	1.5	2.0	2.5	2.5				
0.079.........	1.5	1.5	2.0	2.5	2.5	2.5			
0.109.........			1.5	2.0	2.0	2.0	2.5		
0.138.........				2.0	2.0	2.0	2.0	2.5	
0.168.........				2.0	1.5	2.0	2.0	2.0	2.5

AIRCRAFT WHEEL LOAD–110,000 lb. dual to 200,000 lb. dual; 190,000 lb. dt. to 350,000 lb. dt.; up to 750,000 lb. ddt.

Metal thickness (in.)	Pipe diameter (in.)								
	12	18	24	36	48	60	72	84	96
0.052.........	2.0	2.5	3.0	3.0					
0.064.........	2.0	2.5	2.5	3.0	3.0				
0.079.........	2.0	2.0	2.5	2.5	2.5	3.0			
0.109.........			2.0	2.5	2.5	2.5	3.0		
0.138.........				2.0	2.0	2.5	3.0	3.0	
0.168.........				2.0	2.0	2.5	3.0	3.0	3.0

AIRCRAFT WHEEL LOAD–Up to 1,500,000 lb.

Metal thickness (in.)	Pipe diameter (in.)								
	12	18	24	36	48	60	72	84	96
0.052.........	2.5	2.5	3.0	3.0					
0.064.........	2.5	2.5	2.5	3.0	3.0				
0.079.........	2.5	2.5	2.5	2.5	2.5	3.0			
0.109.........			2.5	2.5	2.5	2.5	3.0		
0.138.........				2.5	2.5	2.5	3.0	3.0	
0.168.........					2.5	2.5	3.0	3.0	3.0

CORRUGATED STEEL 3" × 1" CORRUGATIONS

AIRCRAFT WHEEL LOAD–Up to 30,000 lb. single and up to 40,000 lb. dual

Metal thickness (in.)	Pipe diameter (in.)							
	36	48	60	72	84	96	108	120
0.052..................	1.5	2.0	2.0	2.0				
0.064..................	1.0	1.5	1.5	2.0	2.0	2.0		
0.079..................	1.0	1.0	1.5	1.5	2.0	2.0	2.0	
0.109..................	1.0	1.0	1.0	1.0	1.5	1.5	2.0	2.0
0.138..................	1.0	1.0	1.0	1.0	1.0	1.5	2.0	2.0
0.168..................	1.0	1.0	1.0	1.0	1.0	1.5	2.0	2.0

AIRCRAFT WHEEL LOAD–40,000 lb. dual to 110,000 lb. dual

Metal thickness (in.)	Pipe diameter (in.)							
	36	48	60	72	84	96	108	120
0.052..................	2.5	3.0	3.0	3.0				
0.064..................	2.0	2.5	2.5	3.0	3.0	3.0		
0.079..................	1.5	2.0	2.5	2.5	3.0	3.0	3.0	
0.109..................	1.5	1.5	2.0	2.0	2.0	2.5	3.0	3.0
0.138..................	1.5	1.5	1.5	2.0	2.0	2.0	2.5	2.5
0.168..................	1.5	1.5	1.5	1.5	2.0	2.0	2.0	2.5

AIRCRAFT WHEEL LOAD–110,000 lb. dual to 200,000 lb. dual; 190,000 lb. dt. to 350,000 lb. dt; up to 750,000 lb. ddt.

Metal thickness (in.)	Pipe diameter (in.)							
	36	48	60	72	84	96	108	120
0.052..................	3.0	3.5	3.5					
0.064..................	2.5	3.0	3.5	3.5	3.5			
0.079..................	2.0	2.5	3.0	3.0	3.5	3.5		
0.109..................	2.0	2.0	2.5	2.5	3.0	3.5	3.5	3.5
0.138..................	2.0	2.0	2.0	2.5	3.0	3.0	3.5	3.5
0.168..................	2.0	2.0	2.0	2.0	2.5	2.5	3.0	3.0

AIRCRAFT WHEEL LOAD–Up to 1,500,000 lb.

Metal thickness (in.)	Pipe diameter (in.)							
	36	48	60	72	84	96	108	120
0.052..................	3.0	3.5	3.5					
0.064..................	2.5	3.0	3.5	3.5	3.5			
0.079..................	2.5	2.5	3.0	3.0	3.5	3.5		
0.109..................	2.5	2.5	2.5	2.5	3.0	3.5	3.5	3.5
0.138..................	2.5	2.5	2.5	2.5	3.0	3.0	3.5	3.5
0.168..................	2.5	2.5	2.5	2.5	2.5	2.5	3.0	3.0

STRUCTURAL PLATE PIPE–9" × 2 1/2" CORR. FOR ALUMINUM; 6" × 2" CORRUGATIONS FOR STEEL

AIRCRAFT WHEEL LOAD–Up to 30,000 lb. s. or 40,000 lb. d.	AIRCRAFT WHEEL LOAD–40,000 lb. d. to 110,000 lb. d.	AIRCRAFT WHEEL LOAD–110 k.d. to 200 k.d.; 190 k.d.t. to 350 k.d.t.; to 750 k.d.d.t.	AIRCRAFT WHEEL LOAD–Up to 1,500,000 lb.
Pipe dia. ÷8 but not, less than 1.0'	Pipe dia. ÷6 but not, less than 1.5'	Pipe dia. ÷5 but not, less than 2.0'	Pipe dia. ÷4 but not, less than 2.5'

TABLE 4.15 (Cont.)

NONREINFORCED CONCRETE

AIRCRAFT WHEEL LOAD–Up to 30,000 lb. single and up to 40,000 lb. dual

Pipe type	Pipe diameter (in.) 4	6	8	10	12	15	18	21	24
Std. strength	2.0	2.0	2.0	2.0	2.5	2.5	2.5	2.5	2.5
Extra strength	1.0	1.0	1.5	1.5	1.5	1.5	1.5	1.5	1.5

AIRCRAFT WHEEL LOAD–40,000 lb. dual to 110,000 lb. dual

Pipe type	Pipe diameter (in.) 4	6	8	10	12	15	18	21	24
Std. strength	3.5	4.0	4.0	4.5	5.5	6.0	6.0	6.0	6.0
Extra strength	1.5	2.0	2.5	3.0	3.5	3.5	3.5	3.5	3.5

REINFORCED CONCRETE

AIRCRAFT WHEEL LOAD–Up to 30,000 lb. single and up to 40,000 lb. dual

Reinf. concrete 0.01" crack D-load	Pipe diameter (in.) 12	15	18	21	24	27	30	33	36	42	48	54	60	72	84	96	108	120	132	144
800													1.0	1.0	1.0	1.0	1.0	1.0	1.0	1.0
1000	2.0	2.0	2.0	2.0	2.0	2.0	2.0	2.0	1.5	1.5	1.5	1.0	1.0	1.0	1.0	1.0	1.0	1.0	1.0	1.0
1350	1.5	1.5	1.5	1.5	1.5	1.5	1.5	1.5	1.5	1.0	1.0	1.0	1.0	1.0	1.0	1.0	1.0	1.0	1.0	1.0
2000	1.0	1.0	1.0	1.0	1.0	1.0	1.0	1.0	1.0	1.0	1.0	1.0	1.0	1.0	1.0	1.0	1.0	1.0	1.0	1.0
3000	1.0	1.0	1.0	1.0	1.0	1.0	1.0	1.0	1.0	1.0	1.0	1.0	1.0	1.0	1.0	1.0	1.0	1.0	1.0	1.0

AIRCRAFT WHEEL LOAD–40,000 lb. dual to 110,000 lb. dual

Reinf. concrete 0.01" crack D-load	Pipe diameter (in.) 12	15	18	21	24	27	30	33	36	42	48	54	60	72	84	96	108	120	132	144
800													6.5	5.5	4.5	3.5	2.0	1.5	1.5	1.0
1000	5.5	5.5	5.5	5.5	5.5	5.0	5.0	5.0	4.5	4.5	4.0	4.0	3.5	3.0	2.0	1.5	1.0	1.0	1.0	1.0
1350	4.0	4.0	4.0	4.0	3.5	3.5	3.5	3.5	3.0	3.0	2.5	2.0	2.0	1.5	1.0	1.0	1.0	1.0	1.0	1.0
2000	3.0	3.0	2.5	2.5	2.5	2.0	2.0	2.0	1.5	1.5	1.5	1.0	1.0	1.0	1.0	1.0	1.0	1.0	1.0	1.0
3000	2.0	2.0	1.5	1.5	1.5	1.5	1.5	1.5	1.0	1.0	1.0	1.0	1.0	1.0	1.0	1.0	1.0	1.0	1.0	1.0

AIRCRAFT WHEEL LOAD–110,000 lb. dual to 200,000 lb. dual; 190,000 lb. dual tandem to 350,000 lb. dual tandem; up to 750,000 lb. d.d.t.

Reinf. concrete 0.01" crack D-load	Pipe diameter (in.) 12	15	18	21	24	27	30	33	36	42	48	54	60	72	84	96	108	120	132	144
800																				
1000																				
1350	7.0	7.0	7.0	7.0	7.0	6.5	6.5	6.5	6.0	6.0	6.0	6.0	6.0	6.0	6.0	5.5	5.5	5.0	4.5	4.0
2000	4.0	4.0	4.0	4.0	4.0	4.0	3.5	3.5	3.5	3.5	3.0	2.5	2.0	2.0	2.5	2.5	2.0	2.0	2.0	1.5
3000	3.0	3.0	2.5	2.5	2.0	2.0	2.0	2.0	2.0	1.5	1.5	1.0	1.0	1.0	1.5	1.5	1.0	1.0	1.0	1.0

AIRCRAFT WHEEL LOAD–Up to 1,500,000 lb.

Reinf. concrete 0.01" crack D-load	Pipe diameter (in.) 12	15	18	21	24	27	30	33	36	42	48	54	60	72	84	96	108	120	132	144
2000	7.0	7.0	7.0	7.0	7.0	6.5	6.5	6.5	6.0	6.0	6.0	6.0	6.0	6.0	6.0	6.0	6.0	6.0	6.0	6.0
3000	4.0	4.0	4.0	4.0	4.0	4.0	3.5	3.5	3.5	3.5	3.0	3.0	3.0	3.0	3.0	3.0	3.0	3.0	3.0	3.0

1. Cover depths are measured from top of flexible pavement, however, provide at least 1 foot between bottom of pavement structure and top of pipe.
2. The types of pipe shown are availbale in intermediate sizes, such as 6", 8", 15", 27", 33", etc.
3. For pipe installation in turfed areas use cover depths shown for 30,000 pound single; 40,000 pound dual.
4. Cover depths shown do not provide for freezing conditions. Usually the pipe invert should be below maximum frost penetration.
5. Blanks in tables indicate that pipe will not meet strength requirements.
6. Minimum cover depths shown for flexible pipe are based on use of excellent backfill.
7. Minimum cover depths shown for rigid pipe are based on use of class B bedding.
8. Minimum cover requirements for concrete arch or elliptical pipe may be taken from tables for reinforced concrete circular pipe, providing the outside horizontal span of the arch or elliptical pipe is matched to outside diameter of the circular pipe (assumes that classes of the pipes are the same).
9. Pipe cover requirements for "up to 1,500,000 pounds" are theoretical as gear configuration is not known.

RIGID PAVEMENT

For all types and sizes of pipe use 1.5 foot as minimum cover under rigid pavement (measure from bottom of slab, providing pipe is kept below subbase course). Rigid pipe for loads categorized as "up to 1,500,000 lb." must, however, be either class IV or class V reinforced concrete.

Source: FAA, 2013.

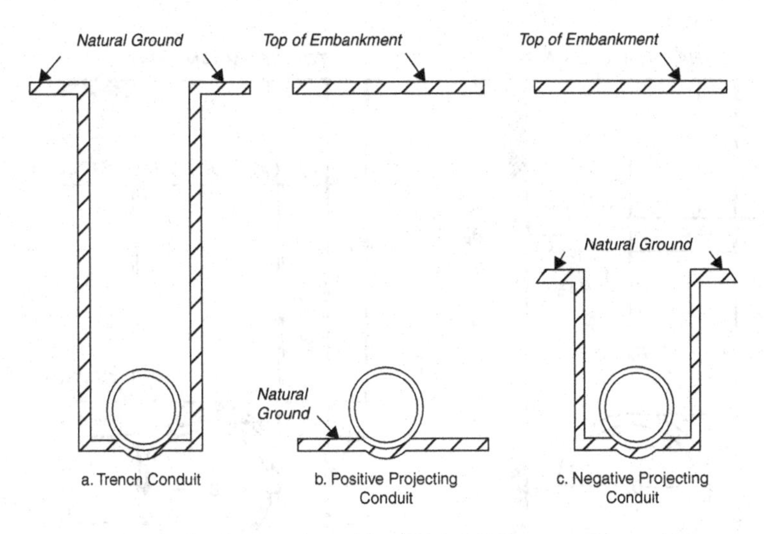

FIGURE 4.26 Three main classes of conduits (FAA, 2013).

For sands, gravelly sands, silty sands, and clayey sands (less than 50% passing 0.075 mm), B is a function of C_u.

For

$$\begin{aligned} &C_u \leq 2 \; or \geq 8, && B=1 \\ &2 \leq C_u \leq 4, && B=0.5C_u \\ &4 < C_u < 8, && B=\frac{8}{C_u} \end{aligned} \tag{4.18}$$

Where,

$$C_u = \frac{D_{60}}{D_{10}} \tag{4.19}$$

For silts and clays, B is a function of the type of geotextile:

For woven geotextiles, B = 1; $O_{95} \leq D_{85}$
For nonwoven geotextiles, B = 1.8; $O_{95} \leq 1.8\ D_{85}$
And for both, AOS or $O_{95} \leq 0.3$ mm

4.7.3 Design of Edgedrains

Guidelines from FHWA (1992) can be used for the design of edgedrains. Generally, the edgedrains are designed to handle the peak flow coming from the permeable base:

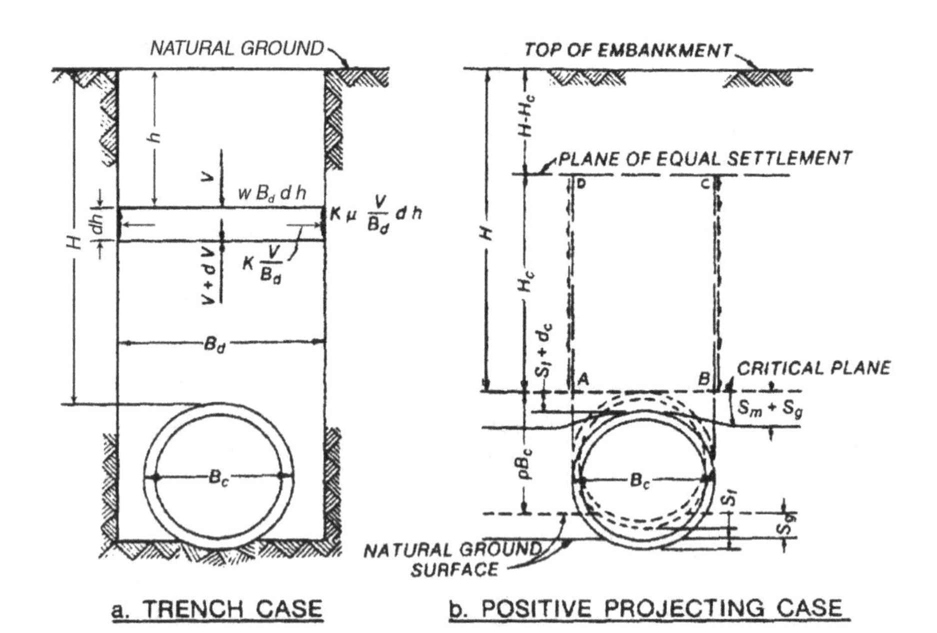

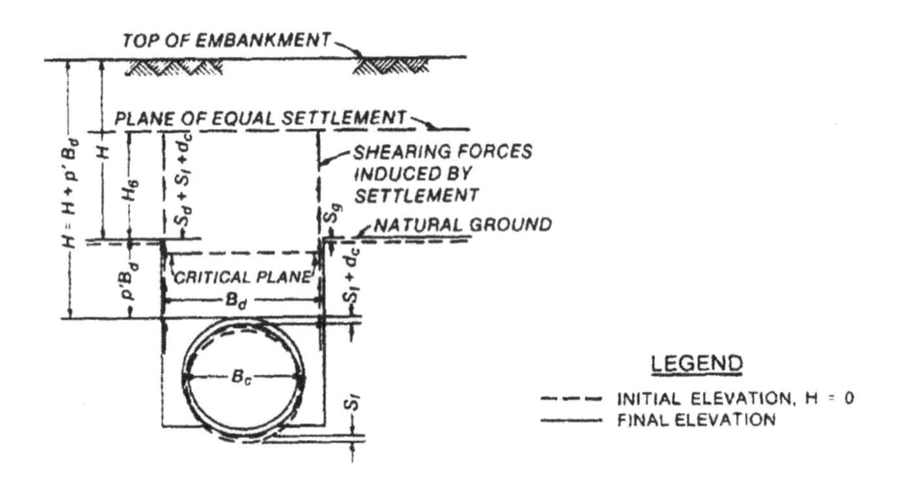

FIGURE 4.27 Free-body conduit diagrams (FAA, 2013).

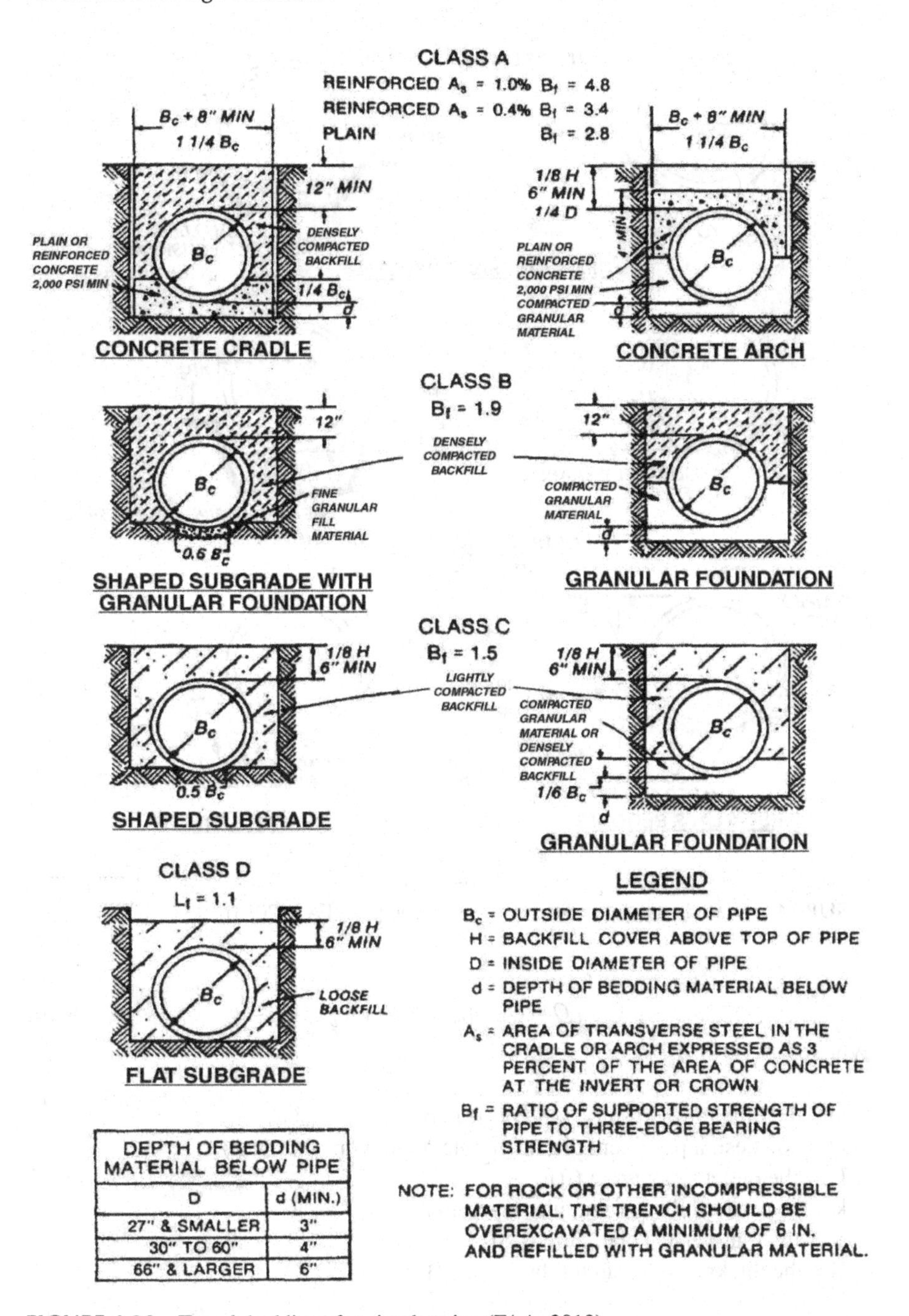

DEPTH OF BEDDING MATERIAL BELOW PIPE	
D	d (MIN.)
27" & SMALLER	3"
30" TO 60"	4"
66" & LARGER	6"

FIGURE 4.28 Trench beddings for circular pipe (FAA, 2013).

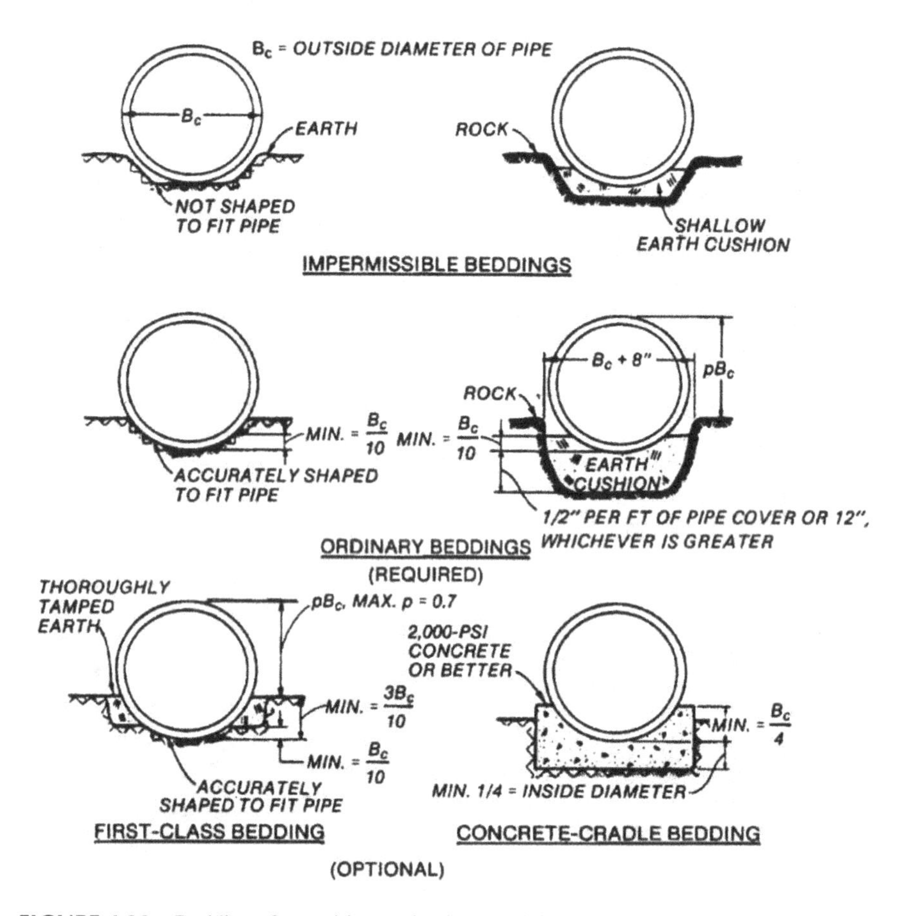

FIGURE 4.29 Beddings for positive projecting conduits (FAA, 2013).

$$Q = Q_p * L_o = (kS_x H)L_o \tag{4.20}$$

Where,

Q = the pipe flow capacity, m^3/day (ft^3/day)
Q_p = the design pavement discharge rate, m^3/day/m (ft^3/day/ft)
L_o = the outlet spacing, m (ft)
k = the permeability of granular layer, m^3/day (ft^3/day)
S_x = the transverse slope, m/m (ft/ft)
H = the thickness of granular layer, m (ft)

For design based on the pavement infiltration flow rate, the design flow capacity of the edgedrain is as follows:

$$Q = Q_p * L_o = (q_i W)L_o \tag{4.21}$$

Where,

q_i = the pavement infiltration, m^3/day/m^2 (ft^3/day/ft^2)
W = the width of the granular layer, m (ft)

For design based on average flow to drain the permeable base, the design flow capacity is as follows:

$$Q = Q_p * L_o = (WHn_eU(\frac{24}{t}))L_o \tag{4.22}$$

Where,

n_e = the effective porosity of granular material
U = the percentage drainage (as 1% = 0.01)
t = the time for drainage, U, to be reached, days

The pipe for the edgedrain is designed according to the Manning equation (Daugherty and Ingersoll, 1954), given earlier as Equation 4.1.

For circular pipe,

$$Q = \frac{K}{n} S^{1/2} (\frac{D}{4})^{2/3} \pi (\frac{D}{2})^2 \tag{4.23}$$

Where,

Q = in m^3/s
D = in m
K = in ($m^{1/3}$)/s

In English units:

$$Q = \frac{53.01}{n} S^{1/2} D^{8/3} \tag{4.24}$$

Where,

Q = in ft^3/day
D = in inches (3–4 in., generally)

The equations are generally used for determining the spacing of the pipes, by fixing a specific type of pipe and with a specific diameter. This is done by setting the pipe capacity equal to the discharge from the unit length of the pavement times the distance between the pipe outlets (spacing).

For infiltration flow:

$$L_o = \frac{Q}{(q_i W)} \tag{4.25}$$

For peak flow:

$$L_o = \frac{Q}{kS_x H} \tag{4.26}$$

For average flow:

$$L_o = \frac{(Qt)}{(24WHn_e U)} \tag{4.27}$$

The spacing is generally used as 250–300 ft. Pipe diameters and spacing are usually determined also in consideration of maintenance requirements.

A filter layer in the form of geotextiles (or, less commonly, aggregates of different gradations) could be wrapped around the pipe or as an envelope to the edgedrains to prevent the inflow of adjacent soil into the pipes but allow the free flow of water into them, and if slotted pipes are used, the filter material must be such that it does not enter the pipes.

Koerner and Hwu (1991) recommend requirements of prefabricated edgedrain filters as follows:

Requirement	Method	Value
Core strength	GRI GG4	≥9600 lbf/in.^{2}
Core flow rate	ASTM D4716	≥15 gal/min–ft
Geotextile permeability	ASTM D4491	≥0.001 cm/s
Geotextile apparent opening size (AOS)	ASTM D4751	≥No. 100 sieve
Geotextile puncture	ASTM D3787	≥75 lb
Geotextile grab strength	ASTM D4632	≥180 lb
Geotextile tear strength	ASTM D4533	≥75 lb

Recommendations (NCHRP, 1994) are that geocomposite edgedrains should be placed on the shoulder side, and the pavement side should be backfilled with suitable sand. The outlet drain pipe is selected so as to make sure that the capacity is equal to, or greater than, the capacity of the edgedrain.

Recommendations (FHWA, 1989) regarding edgedrains and outlets are as follows:

1. The preferable location of the edgedrain is under the shoulder just adjacent to the pavement/shoulder joint, with the top of the pipe at the bottom of the layer to be drained, in a 12 inch trench.

2. The filter fabric could be provided at the subbase–edgedrain interface to prevent clogging of the filter by fines, and the trench backfill material should have adequate permeability.
3. Outlet spacing should be less than 500 ft, with additional outlets at the bottom of vertical sags, with rigid PVC outlets being desirable along with headwalls to protect it, prevent erosion, and help in locating the outlets.

4.8 USE OF SOFTWARE FOR THE DESIGN OF DRAINAGE STRUCTURES

Figure 4.30 explains the step-by-step procedures for the use of Drainage Requirement in Pavements (DRIP) software for the design of drainage system for an example pavement (DRIP User's Guide, FHWA, 2002); Figure 4.31 shows the screenshots for the example. DRIP, Version 2 can be downloaded from: www.me-design.com/MEDesign/DRIP.html?AspxAutoDetectCookieSupport=1

PROBLEM

A pavement section consists of two 12 ft (3.66 m) lanes of 9 in. (225 mm) thick PCC pavement with 10-ft (3.05-m) AC shoulders on each side with a uniform cross slope (not crowned), and the width of the permeable base is the same as the PCC pavement. The transverse joint spacing is 20 ft (6.1 m). The slope in both the longitudinal (S) and transverse (S_x) directions is 2%. The permeable base is made up of AASHTO #57 material and has a unit weight of 100 pcf (1600 kg/m3), specific gravity of 2.65, and a minimum permeability of 3000 ft/day (914 m/day). The thickness of the permeable base is 4 in. (100 mm), based on construction considerations. Assume a unit weight of 162 pcf (2595 kg/m3) for the PCC material. The subgrade is Georgia red clay, which is actually a well-graded clayey-silt. Laboratory tests indicate the particle gradation for the subgrade material as follows: 92%, 67%, 55%, 42% and 31% passing the No. 4, 10, 20, 50 and 200 sieve, respectively, and a permeability of 0.0033 ft/day (0.001 m/day). Corrugated pipe edgedrains having 4 in. (100-mm) diameter are used on the project.

SOLUTION

Step 1: Select *Unit* (English), and *Mode* (normal), *Sensitivity* (length, slope, permeability, inflow, drain, thickness, porosity), *Plot Scale* (power 0.45) (Figure 4.31a).

Step 2: Enter information for roadway geometry (Figure 4.31b).
Enter the value of b as 24 ft (2 lanes of 12 ft each). Since the problem statement assumes that the width of the permeable base is the same as that of the PCC layer, enter the value of c as 0. Clicking the calculator icon for width of the permeable base W results in a computed value of 24 ft. Enter the longitudinal and transverse slopes in the S and SX edit boxes, respectively. Click the calculator icons for both SR (resultant slope) and LR (resultant length) to compute these parameters. The calculations should yield the following values: SR = 0.0283 ft/ft and LR = 33.94 ft/ft. Select a uniformly cross-sloped pavement section by clicking on the Geometry B radio button.

Step 3: Enter information for inflow.
Click on the *Crack Infiltration Method* radio button (since this method was required to be chosen to estimate inflow). Retain the default values of 2.4 ft3/day/feet for Ic and 0 for pavement permeability kp. Also retain the values for Wc and W that have been automatically carried forward from the calculations performed on the Roadway Geometry page. Enter the number of longitudinal cracks, Nc, as 3 (Nc = the number of contributing lanes + 1 = 2 + 1 = 3). Enter the given transverse spacing of contributing transverse joints Cs = 20 ft. Click the calculator icon to compute inflow qi. This should yield a value of 0.42ft3/day/ft2.

FIGURE 4.30 Flowchart for solving example problem with DRIP (continued on pages 105–106).

Step 4: *Continuing information for Inflow* – meltwater computation
Activate the *Meltwater* sub-screen in the *Inflow* property page by clicking on the *Include Meltwater in the inflow calculations* checkbox. The heave rate (*Heave*) for the clayey-silt subgrade soil can be determined by clicking on the calculator button located to the left of the variable. This action calls up the heave rate table (Figure 4.31c). The heave rate table shows that the value for clayey silt soil is 0.51 in./day. This value can either be entered manually in the edit box on the *Meltwater* sub-screen or entered automatically by selecting the row corresponding to clayey-silt soil type. For materials with large heave rates, the mid-point values may be entered. Enter the given values for the subgrade permeability (ksub = 0.0033 ft/day), unit weight of the pavement surface (γp = 162 lb/ft3), the unit weight of the base material above the subgrade (γb = 100 lb/ft3), the thickness of pavement surface (Hs = 9 in, 0.75 ft.), and the thickness of the base material (H = 4 in., 0.33 ft) in the appropriate boxes. Click on the calculator icon next to the variable σ to compute the stress imposed by the pavement on the subgrade. This will yield a value of 154.5 lb/ft2. Click the calculator icon for determining the quantity of meltwater (q_m).

Moulton's meltwater chart appears on the screen (Figure 4.31d). On this screen a horizontal redline appears marking the heave rate of the subgrade (in mm/day for 0.51 in./day). Use the mouse to slide the vertical tick line along the horizontal red line (0.51 in./day line) to an approximate location for a stress of 154.8 psf. The value for q_m, 0.05 will appear in the appropriate box. When satisfied that the mouse is positioned at the proper stress value, double click the left mouse button to produce a computed value of q_m (0.05ft3/day/ft^2).

Step 5: Enter information for the permeable base (design using depth of flow method):
Select the *Permeable Base* tab to access the corresponding property page. Click on the *Depth of Flow Method* radio button. Most of the variables such as qi + qm, SR, LR and H already have values carried forward from pervious screens. The term qi + qm is a summation of the inflow from rainfall and meltwater. Enter the coefficient of permeability of the base (k = 3000 ft/day). Compute the Hmin by clicking on the calculator icon. A Hmin value of 0.1437 ft is reported, which is lower than the selected permeable base thickness (H) of 0.3333 ft. Therefore, the design is satisfied. If Hmin were much greater than H, the designer would make adjustments to the design by changing design variables such as base thickness or permeability. If the base thickness H can be revised, this parameter must be changed to be at least equal to the Hmin. Compute flow capacity of the permeable base (qd) by clicking on the appropriate calculator button. This yields a value of 44 ft3/day/ft. The flow capacity is estimated from Moulton's chart (Figure 4.31e).

Additional discussion on depth of flow design
If the user wishes to perform a sensitivity analysis to see the influence of different parameters on the required thickness, the graph icon to the right of Hmin variable in Figure 4.31f can be clicked. DRIP internally performs the calculations over a pre-selected range of the independent parameter (in this case the required base thickness) and computes and plots the dependent variables. The charts are plotted using the *DripPlot*. The *DripPlot* window appears as soon as the graph icon next to the Hmin variable is clicked and the following sensitivity plots are displayed: (Figure 4.31g)

FIGURE 4.30 Flowchart for solving example problem with DRIP (continued on page 106).

Required Base Thickness (Hmin) vs. Pavement Inflow (qi or qi + qm).
Required Base Thickness (Hmin) vs. Base Course Permeability (k).
Required Base Thickness (Hmin) vs. Resultant Slope (SR).
Required Base Thickness (Hmin) vs. Resultant Length (LR).
The user can control the type of plots to be displayed in the DripPlot window using the *Options / Sensitivity* menu command prior to clicking the graph icon.

Step 6: Enter information for separator layer (design):
Select the *Separator Layer* tab. Select the *No Separator* radio button to evaluate the need for a separator layer. Click on the calculator icon next to the *Base Course* variable. This shifts the user to the *Sieve Analysis* property page (Figure 4.31h). The permeable base is made up of the AASHTO #57 material. The gradation for this material is already in the sieve analysis library. To initialize this gradation to the program, click on the dropdown material library list box and select AASHTO #57. Click on the calculator icon to determine the particle sizes for D15 and D50. Select the *Separator Layer* tab again. The values of D15 = 0.2529 in. and D50 = 0.5484 in. appear for the *Base Course*. Click on the calculator icon next the *Subgrade* variable to compute the subgrade particle sizes. The user is returned to the *Sieve Analysis* property page again and the Subgrade radio button is automatically activated (Figure 4.31i). Select the *Value* radio button and enter the subgrade gradation in the grid on the left-hand side of the property page. Compute the D50 and D85 by clicking on the particle size calculator icon. Return to the separator layer property page by selecting the *Separator* tab. The values of D50 = 0.0253 in. and D85 = 0.1498 in. are returned for the subgrade (Figure 4.31j). On the *Sieve Analysis* property page, check the filtration and uniformity criteria at the subgrade/base interface by clicking on the balance icon on the right-hand side. Both the criteria generate a *Pass* rating, which implies that no separator layer is required (Figure 4.31k).

Step 7: Enter the information for the edgedrain (design):
Select the *Edgedrain* tab. Select the *Pipe Edgedrain* radio button. Click the *Corrugated* checkbox and a value of nmanning = 0.024 is displayed. Type in a value of D = 4 in. The calculator icon is enabled for calculating the pipe capacity (Q). The computed pipe capacity should read 12,594 ft3/day. Click on the *Lo* button to compute the outlet spacing. This gives an Lo value of 391.6 ft. However, this value is based on the default selection of the *Pavement Infiltration* method for the pavement discharge rate. Two other methods are given to compute the maximum outlet spacing: *Permeable Base* and *Time-to-Drain*. Since the time-to-drain method was not used for designing the permeable base we will not examine this approach. For computing the outlet spacing Lo based on the *Permeable Base* approach, select the appropriate radio button from the *Discharge Rate Approach* options. Our previously computed base discharge qd of 44.81 ft3/day/ft appears automatically. This results in a maximum outlet spacing of 281 ft. Thus, the permeable base discharge result is the critical value, so the user should specify an outlet spacing of 281 ft (Figure 4.31l).

Step 8: Reviewthe input and output by clicking on *File → Export Summary* (Figure 4.31m).

FIGURE 4.30 Flowchart for solving example problem with DRIP.

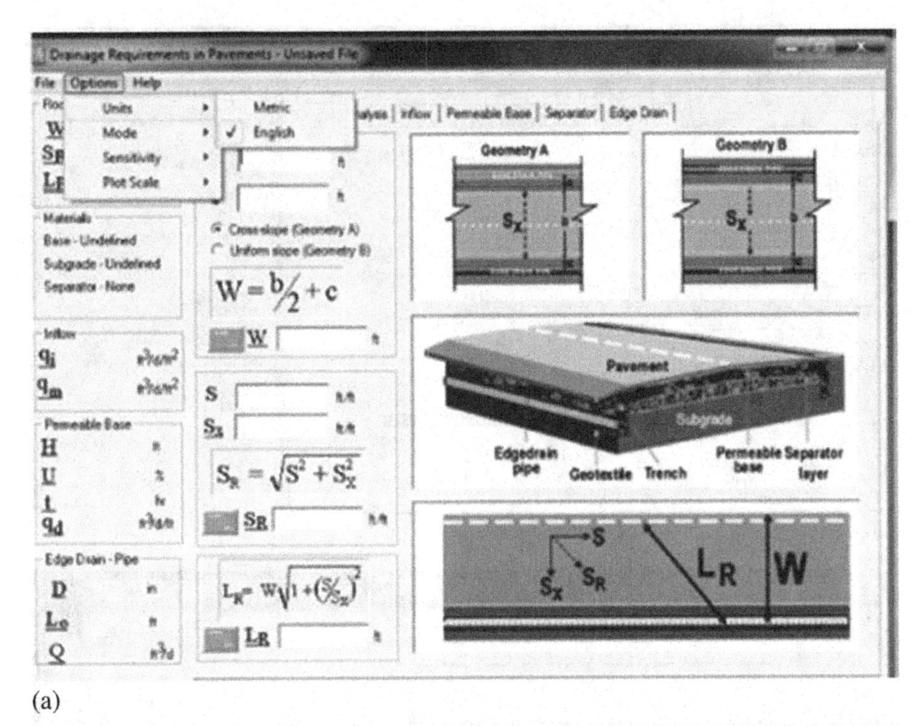

(a)

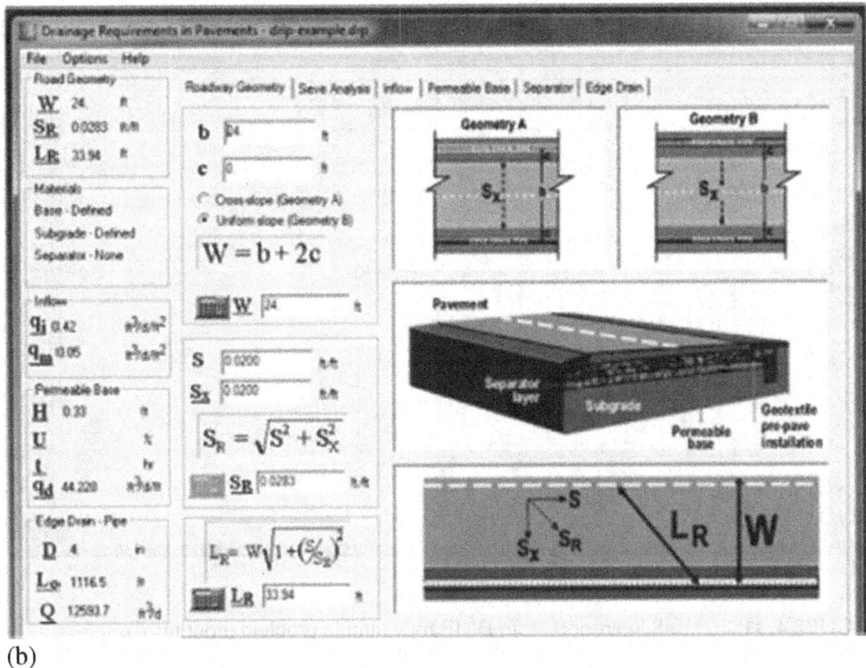

(b)

FIGURE 4.31 (a, b) Screenshots from DRIP for example problem.

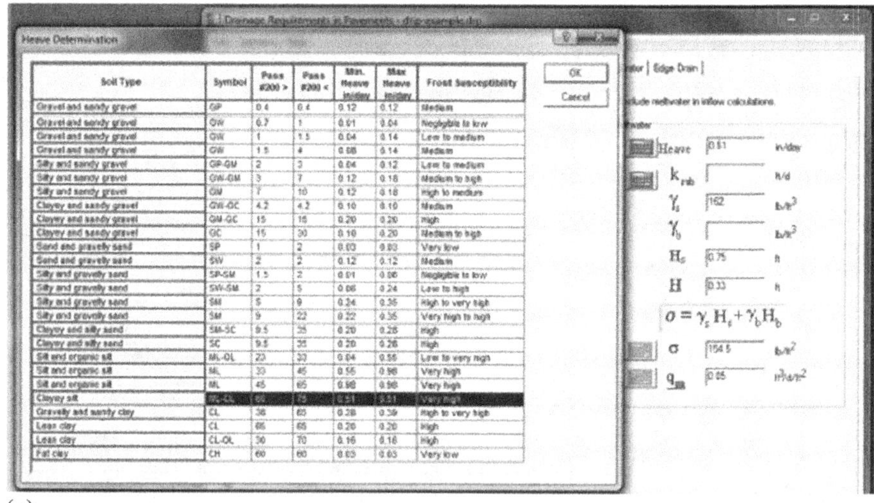

Heave Determination

Soil Type	Symbol	Pass #200 >	Pass #200 <	Min. Heave in/day	Max Heave in/day	Frost Susceptibility
Gravel and sandy gravel	GP	0.4	0.4	0.12	0.12	Medium
Gravel and sandy gravel	GW	0.7	1	0.01	0.04	Negligible to low
Gravel and sandy gravel	GW	1	1.5	0.04	0.14	Low to medium
Gravel and sandy gravel	GW	1.5	4	0.08	0.14	Medium
Silty and sandy gravel	GP-GM	2	3	0.04	0.12	Low to medium
Silty and sandy gravel	GW-GM	3	7	0.12	0.18	Medium to high
Silty and sandy gravel	GM	7	10	0.12	0.18	High to medium
Clayey and sandy gravel	GW-GC	4.2	4.2	0.10	0.10	Medium
Clayey and sandy gravel	GM-GC	15	15	0.20	0.20	High
Clayey and sandy gravel	GC	15	30	0.10	0.20	Medium to high
Sand and gravelly sand	SP	1	2	0.03	0.03	Very low
Sand and gravelly sand	SW	2	2	0.12	0.12	Medium
Silty and gravelly sand	SP-SM	1.5	2	0.01	0.06	Negligible to low
Silty and gravelly sand	SW-SM	2	5	0.06	0.24	Low to high
Silty and gravelly sand	SM	5	9	0.24	0.35	High to very high
Silty and gravelly sand	SM	9	22	0.22	0.35	Very high to high
Clayey and silty sand	SM-SC	9.5	35	0.20	0.28	High
Clayey and silty sand	SC	9.5	35	0.20	0.28	High
Silt and organic silt	ML-OL	23	33	0.04	0.56	Low to very high
Silt and organic silt	ML	33	45	0.56	0.98	Very high
Silt and organic silt	ML	45	65	0.98	0.98	Very high
Clayey silt	ML-CL	60	75	0.51	0.51	Very high
Gravelly and sandy clay	CL	36	65	0.28	0.39	High to very high
Lean clay	CL	65	65	0.20	0.20	High
Lean clay	CL-OL	30	70	0.16	0.16	High
Fat clay	CH	60	60	0.03	0.03	Very low

(c)

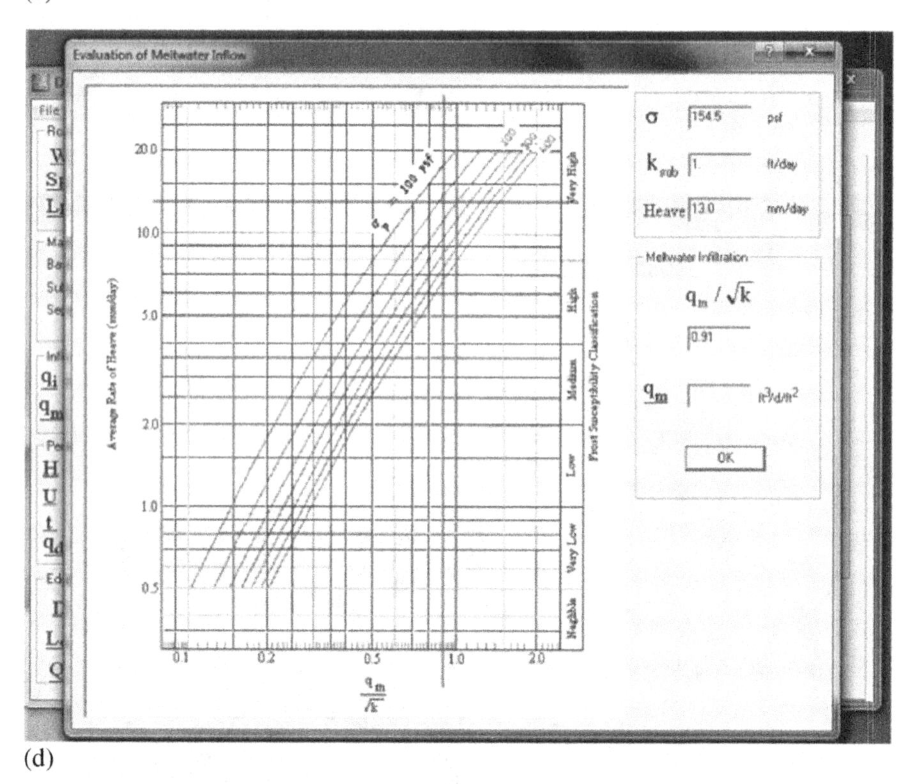

(d)

FIGURE 4.31 (c, d) Screenshots from DRIP for example problem (continued).

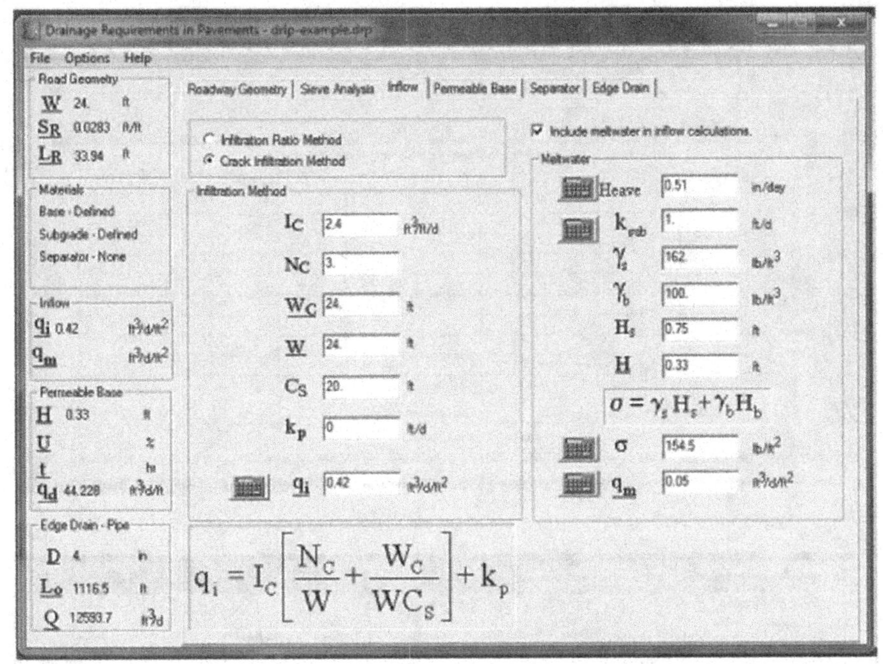

(e)

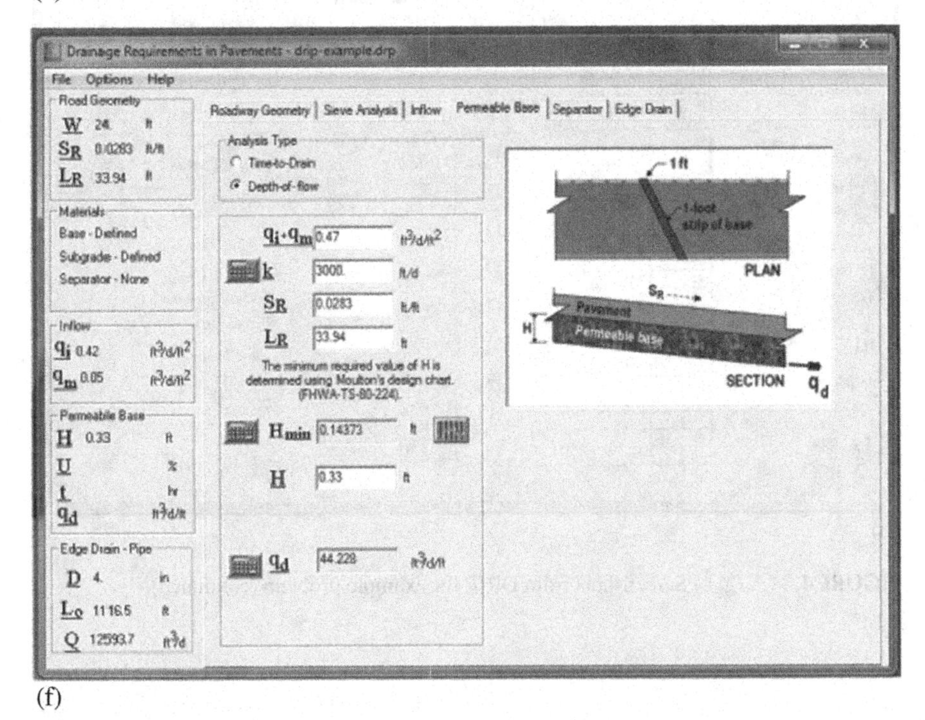

(f)

FIGURE 4.31 (e, f) Screenshots from DRIP for example problem (continued).

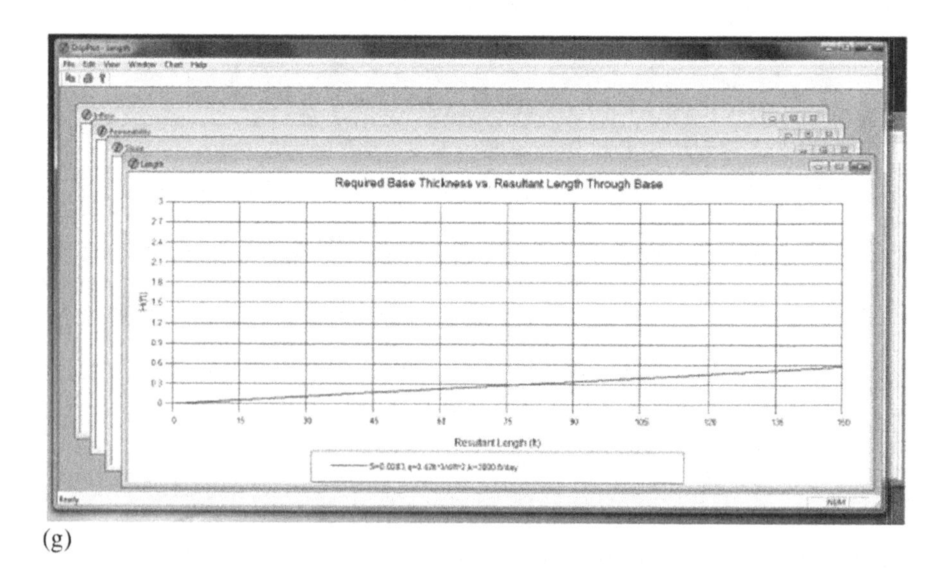

(g)

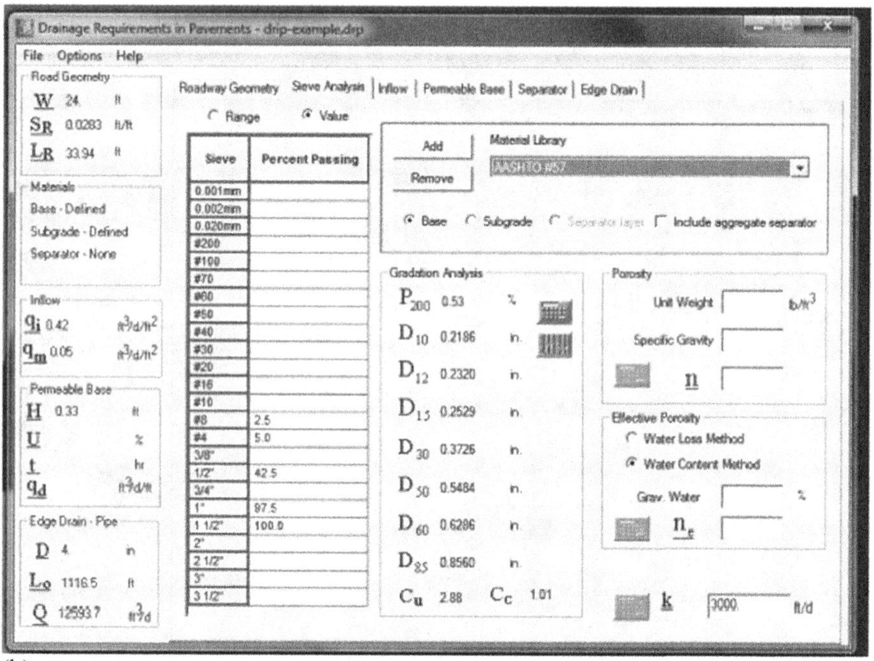

Drainage Requirements in Pavements - drip-example.drp

File Options Help

Road Geometry
W 24 ft
S_R 0.0283 ft/ft
L_R 33.94 ft

Materials
Base - Defined
Subgrade - Defined
Separator - None

Inflow
q_i 0.42 $ft^3/d/ft^2$
q_m 0.05 $ft^3/d/ft^2$

Permeable Base
H 0.33 ft
U %
t hr
q_d $ft^3/d/ft$

Edge Drain - Pipe
D 4 in
L_o 1116.5 ft
Q 12593.7 ft^3/d

Roadway Geometry | Sieve Analysis | Inflow | Permeable Base | Separator | Edge Drain

Range Value

Sieve	Percent Passing
0.001mm	
0.002mm	
0.020mm	
#200	
#100	
#70	
#60	
#50	
#40	
#30	
#20	
#16	
#10	
#8	2.5
#4	5.0
3/8"	
1/2"	42.5
3/4"	
1"	97.5
1 1/2"	100.0
2"	
2 1/2"	
3"	
3 1/2"	

Add
Remove
Material Library
AASHTO #57
Base Subgrade Separator layer Include aggregate separator

Gradation Analysis
P_{200} 0.53 %
D_{10} 0.2186 in.
D_{12} 0.2320 in.
D_{15} 0.2529 in.
D_{30} 0.3726 in.
D_{50} 0.5484 in.
D_{60} 0.6286 in.
D_{85} 0.8560 in.
C_u 2.88 C_c 1.01

Porosity
Unit Weight lb/ft^3
Specific Gravity
n

Effective Porosity
Water Loss Method
Water Content Method
Grav. Water %
n_e

k 3000 ft/d

(h)

FIGURE 4.31 (g, h) Screenshots from DRIP for example problem (continued).

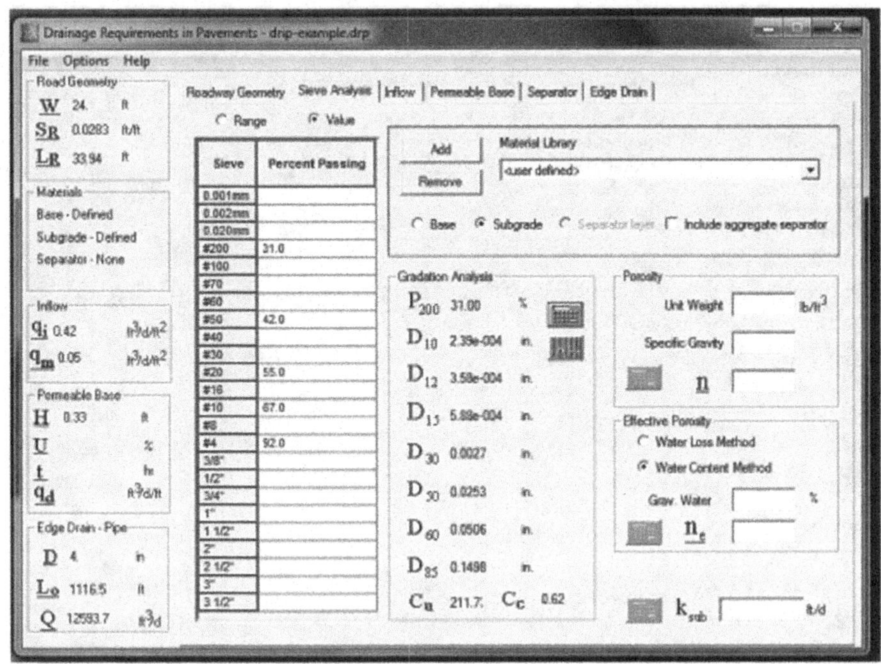

(i)

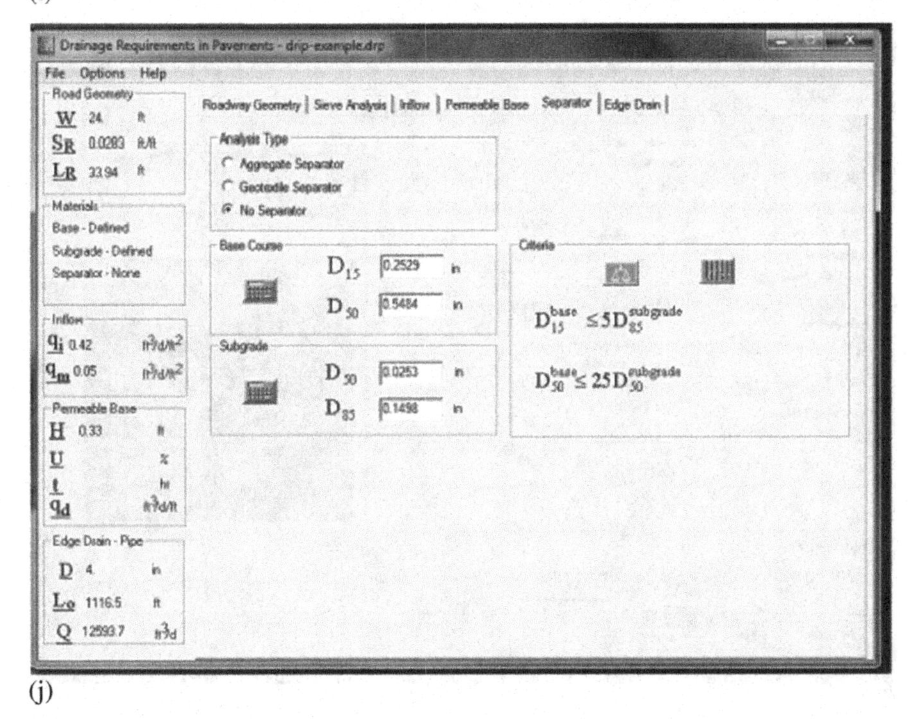

(j)

FIGURE 4.31 (i, j) Screenshots from DRIP for example problem (continued).

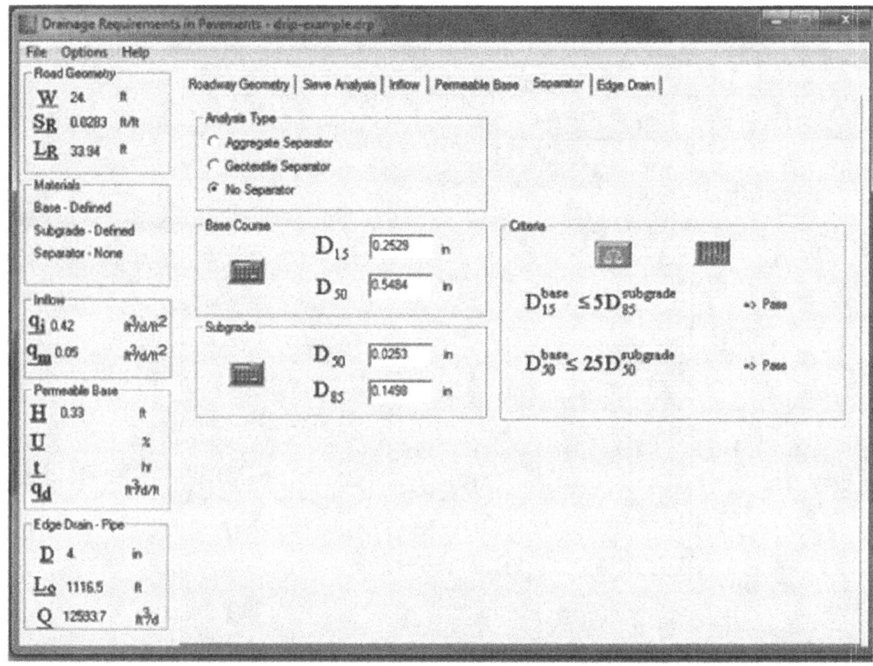

(k)

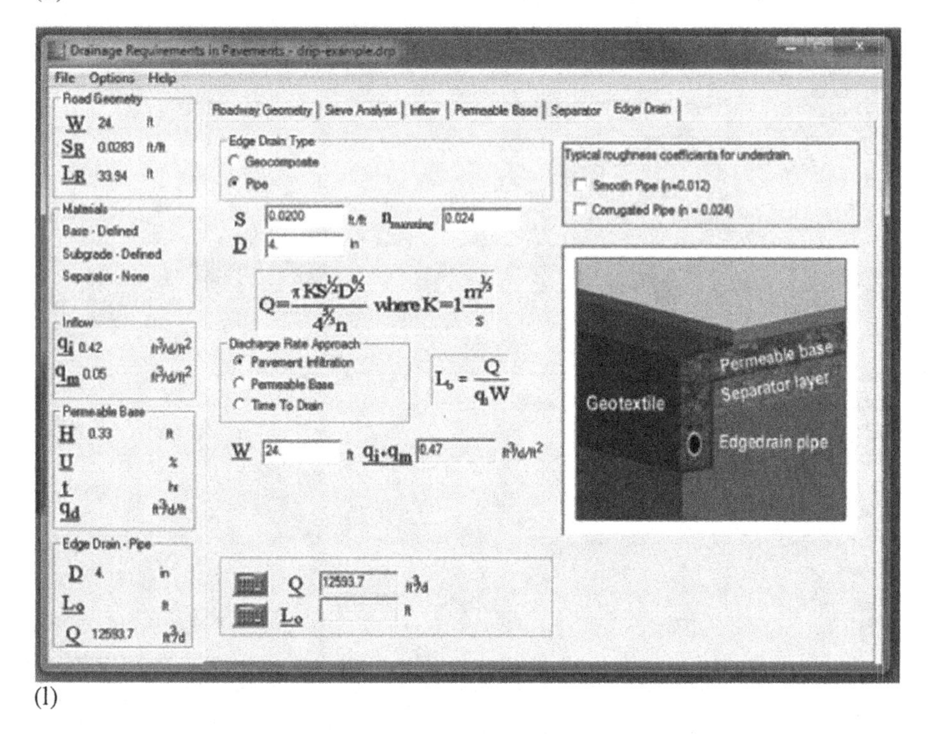

(l)

FIGURE 4.31 (k, l) Screenshots from DRIP for example problem (continued).

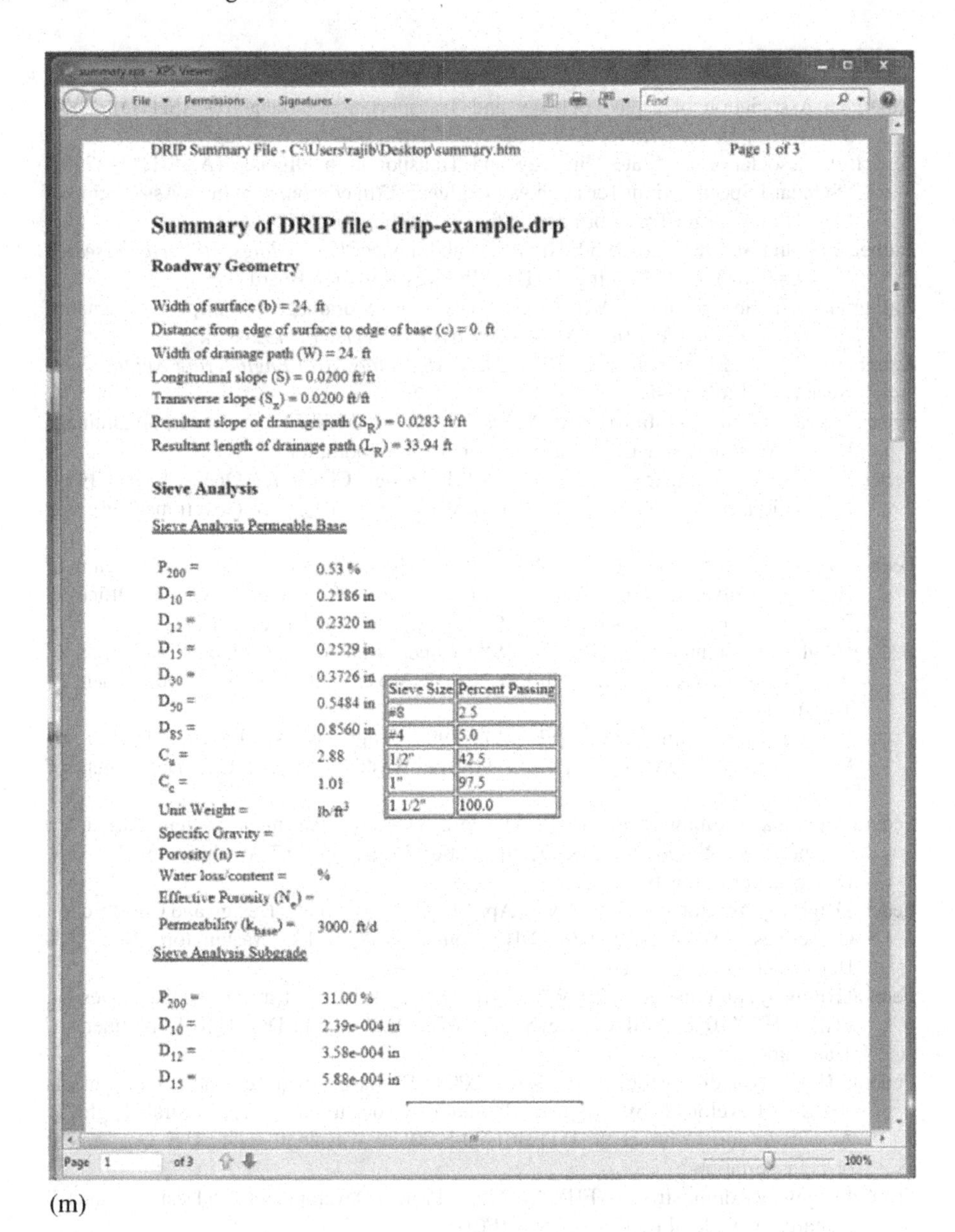

Summary of DRIP file - drip-example.drp

Roadway Geometry

Width of surface (b) = 24. ft
Distance from edge of surface to edge of base (c) = 0. ft
Width of drainage path (W) = 24. ft
Longitudinal slope (S) = 0.0200 ft/ft
Transverse slope (S_x) = 0.0200 ft/ft
Resultant slope of drainage path (S_R) = 0.0283 ft/ft
Resultant length of drainage path (L_R) = 33.94 ft

Sieve Analysis

Sieve Analysis Permeable Base

P_{200} = 0.53 %
D_{10} = 0.2186 in
D_{12} = 0.2320 in
D_{15} = 0.2529 in
D_{30} = 0.3726 in
D_{50} = 0.5484 in
D_{85} = 0.8560 in
C_u = 2.88
C_c = 1.01
Unit Weight = lb/ft³
Specific Gravity =
Porosity (n) =
Water loss/content = %
Effective Porosity (N_e) =
Permeability (k_{base}) = 3000. ft/d

Sieve Size	Percent Passing
#8	2.5
#4	5.0
1/2"	42.5
1"	97.5
1 1/2"	100.0

Sieve Analysis Subgrade

P_{200} = 31.00 %
D_{10} = 2.39e-004 in
D_{12} = 3.58e-004 in
D_{15} = 5.88e-004 in

(m)

FIGURE 4.31 (m) Screenshots from DRIP for example problem (continued).

REFERENCES

American Association of State Highway and Transportation Officials (AASHTO). 2000. *MDM-SI-2, Model Drainage Manual*, 2000 Metric Edition. Washington, DC.

American Association of State Highway and Transportation Officials (AASHTO). 2002. "Standard Specifications for Highway Bridges," 17th edition, American Association of State Highway and Transportation Officials, Washington, DC.

Barber, E.S. and Sawyer, C.L. 1952. Highway subdrainage. *Proceedings, Highway Research Board*, pp. 643–666. Washington, DC: Highway Research Board.

Casagrande, A. and Shannon, W.L. 1952. Base course drainage for airport pavements. *Proceedings of the American Society of Civil Engineers*, 77: 792–814.

Daugherty, R.L. and Ingersoll, A.C. 1954. *Fluid Mechanics with Engineering Applications*. New York: McGraw-Hill.

Federal Aviation Administration (FAA). August 15, 2013. AC 150/5320-5D – Airport Drainage Design. Washington, DC: U.S. Department of Transportation.

Federal Highway Administration (FHWA). 1961. Design Charts for Open-Channel Flow, Hydraulic Design Series No. 3 (HDS 3). Washington, DC: U.S. Government Printing Office.

Federal Highway Administration (FHWA). 1965. "Hydraulic Charts for the Selection of Highway Culverts," HEC No. 5, Hydraulics Branch, Bridge Division, Office of Engineering, FHWA, Washington, D.C. 20590 (L.A. Herr and H.G. Bossy).

Federal Highway Administration (FHWA). 1989. Concrete Pavement Drainage Rehabilitation, FHWA Report, Experimental Project No. 12. Washington, DC: U.S. Department of Transportation.

Federal Highway Administration (FHWA). 1990. Highway Subdrainage Design by Microcomputer: (DAMP), FHWA-IP-90-012. Washington, DC: U.S. Department of Transportation.

Federal Highway Administration (FHWA). 1992. Drainage Pavement System Participant Notebook, FHWA-SA-92-008, Demonstration Project No. 87. Washington, DC: U.S. Department of Transportation.

Federal Highway Administration (FHWA). April 1998. Geosynthetic Design and Construction Guidelines, FHWA-HI-95-038, NHI Course No. 13213. Washington, DC: U.S. Department of Transportation.

Federal Highway Administration (FHWA). April 1999. Pavement Subsurface Drainage Design, FHWA-HI-99-028, NHI Course No. 131026. Washington, DC: U.S. Department of Transportation.

Federal Highway Administration (FHWA). 2002. Drainage Requirements in Pavements (DRIP), Developed by Applied Research Associates for the Federal Highway Administration, Contract No. DTFH61-00-F-00199. Washington, DC: U.S. Department of Transportation.

Federal Highway Administration (FHWA). 2008. "Project Development and Design Manual," prepared by Federal Lands Highway (FLH).

Federal Highway Administration (FHWA). 2009. "Urban Drainage Design Manual," Hydraulic Engineering Circular 22, Report NHI-10-09 (S.A. Brown, J.D. Schall, J.L. Morris, C.L. Doherty, S.M. Stein, J.C. Warner).

Federal Highway Administration (FHWA). 2018. HY-8 Culvert Hydraulic Analysis Program www.fhwa.dot.gov/engineering/hydraulics/software/hy8/.

Federal Highway Administration (FHWA), USDOT. 2010. Culvert Assessment and Decision Making Procedures Manual For Federal Lands Highway Publication No. FHWA-CFL/TD-10-005 September.

Federal Highway Administration U.S. 2012. Department of Transportation Hydraulic Design of Highway Culverts Third Edition Publication No. FHWA-HIF-12–26 April.

Garber, N.J. and Hoel, L.A. 2002. *Traffic and Highway Engineering*, 3rd ed. Clifton Park, NY: Thomson Learning.

Koerner, R.M. and Hwu, B.-L. 1991. Prefabricated Highway Edge Drains, Transportation Research Record No. 1329. Washington, DC: Transportation Research Board.

Laramie County Conservation District. N.d. Best management Practice for Stormwater Runoff. http://repo.floodalliance.net/jspui/bitstream/44111/1348/1/Best%20management%20practices%20for%20stormwater%20runoff.pdf.

Moulton, L.K. 1979. Design of Subsurface Drainage Systems for Control of Groundwater, Transportation Research Record No. 733. Washington, DC: Transportation Research Board.

Moulton, L.K. 1980. Highway Subsurface Design, FHWA-TS-80–224. Washington, DC: U.S. Department of Transportation.

National Cooperative Highway Research Program (NCHRP). 1994. Long-term Performance of Geosynthetics in Drainage Applications, NCHRP Report 367. Washington, DC: Transportation Research Board, National Research Council, National Cooperative Highway Research Program.

National Environmental Protection Agency. N.d. NEPA, Public Law [PL] 91–190 Whole Building Design Guide's (WBDG) www.wbdg.org/.

Unified facilities criteria (UFC) Draft. 2006. Surface Drainage Design, Report No. AC 150/5320-5D, Federal Aviation Administration (FAA), USA.

USFS. 2006. "Low-water Crossings: Geomorphic, Biological and Engineering Design Considerations," U.S. Forest Service Report 0625 1808 SDTDC (K. Clarkin, G. Keller, T. Warhol and S. Hixson).

Vardanega, P.J. 2014. State of the art: permeability of asphalt concrete. *Journal of Materials in Civil Engineering*, 26(1): 54–64.

5 Factors that Govern the Flow of Water

5.1 HYDRAULIC CONDUCTIVITY

Water flows through various layers in a pavement, each of which has certain flow characteristic. The basic property that governs the flow in porous media is hydraulic conductivity. D'Arcy's law states that:

$$v = ki \tag{5.1}$$

Where,

v = the velocity of flow
i = the hydraulic gradient (head loss per unit length of flow)
k = the coefficient of permeability or hydraulic conductivity or simply permeability

Constant head (ASTM D-2434) or falling head tests (ASTM D-4630) can be conducted to estimate k of different pavement materials. Empirical relations to estimate k are also available, as indicated in equations 4.5 and 4.6.

Some of the existing water in the soil remains as thin films around soil particles, and the amount that actually flows is dependent on the effective porosity in a soil as defined in Chapter 4.

Reported values of W_L (see equation 4.13) expressed as a percentage are as follows:

Amount of Fines	<2.5% Fines			5% Fines			10% Fines		
Type of fines	Filler	Sand	Clay	Filler	Sand	Clay	Filler	Sand	Clay
Gravel	70	60	40	60	40	20	40	30	10
Sand	57	50	35	50	35	15	25	18	8

5.2 UNSATURATED HYDRAULIC CONDUCTIVITY

The estimation and prediction of hydraulic conductivity under unsaturated condition has been a topic of research for many years. Under unsaturated conditions soil suction plays an important role and dictates the hydraulic conductivity.

Different researchers have developed methods to relate the suction with the volumetric water content of the soil, typically what is known as the Soil Water Characteristics Curves (SWCC). To avoid testing to generate the SWCC, researchers have developed both closed form solutions as well as those based on the index properties of the soil.

Examples of closed form solutions are from Gardner (1958), Brooks and Corey (1964), van Genuchten (1980), Fredlund and Xing (1994) and Fredlund, Xing and Huang (1994). The expressions are as follows.
Gardner, 1958:

$$\theta = \theta_r + \frac{\theta_s - \theta_r}{1 + \left(\frac{\psi}{\alpha}\right)^n} \tag{5.2}$$

Brooks and Corey, 1964:

$$\theta = \theta_r + (\theta_s - \theta_r)\left(\frac{\psi_b}{\psi}\right)^\lambda \tag{5.3}$$

θ = volumetric water content
θ_s = water content at saturation
θ_r = residual water content
ψ = matric suction
α = value of ψ at a θ of $(\theta_s + \theta_r)/2$
ψ_b = air entry value
λ = pore size distribution index

Van Genuchten, 1980:

$$\theta = \theta_r + \frac{\theta_s - \theta_r}{\left[1 + \left(\frac{\psi}{a}\right)^n\right]^m} \tag{5.4}$$

Fredlund and Xing (1994, 1) :

$$\theta = \theta_r + \frac{\theta_s - \theta_r}{\left\{ ln\left[e + \left(\frac{\psi}{a} \right)^n \right] \right\}^m} \tag{5.5}$$

Fredlund and Xing (1994, 2) :

$$\theta = C(\psi) \frac{\theta_s}{\left\{ ln\left[e + \left(\frac{\psi}{a} \right)^n \right] \right\}^m} \tag{5.6}$$

As an example, the curves that are generated for a specific soil, using the above five methods are shown in Figure 5.1. The corresponding parameters for each model are shown in Table 5.1.

Expressions from Van Genuchten (1980) and Fredlund and Xing (1994) can also be used with soil index properties (different properties for fine and coarse grained soils) to develop SWCC, as shown in Tables 5.2 and 5.3, and Figures 5.2 and 5.3.

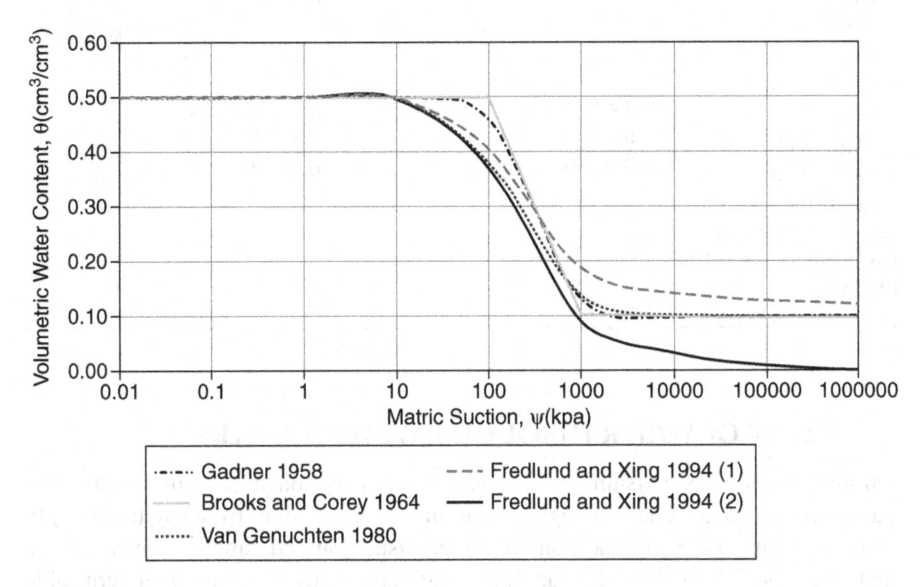

FIGURE 5.1 SWCC from closed form solutions. (Courtesy, Dr. Ghada S. Ellithy, US Army Corps of Engineers. Used with permission.)

TABLE 5.1
Parameters for the Different Models (With Kind Permission From Dr. Ghada S. Ellithy)

Gardner, 1958	Brooks and Corey, 1964	Van Genuchten, 1980	Fredlund and Xing (1994, 1)	Fredlund and Xing (1994, 2)
α = 300 kPa n = 2 θ_r = 0.1 cm3/cm3 θ_s = 0.5 cm3/cm3	ψ_b = 100 kPa λ =2 θ_r = 0.1 cm3/cm3 θ_s = 0.5 cm3/cm3	a = 100 kPa n = 2 m = 1-1/n 0.5 θ_r = 0.1 cm3/cm3 θ_s = 0.5 cm3/cm3	a = 100 kPa n = 2 m = 1 θ_r = 0.1 cm3/cm3 θ_s = 0.5 cm3/cm3	a = 100 kPa n = 2 m = 1 θ_s = 0.5 cm3/cm3

TABLE 5.2
Index Properties and Estimated of Parameters for Fine Grained Soils (With Kind Permission From Dr. Ghada S. Ellithy)

Soil Properties	Parameters for Van Genuchten, 1980 method	Parameters for Fredlund and Xing (1994) method
Void Ratio, e = 1.1 Liquid Limit, LL = 55 % Plastic Limit, PL = 25 % Sand Content (< 2 mm) = 20 % Fines Content (< 0.075 mm) = 70 % Clay Fraction (< 0.002 mm) = 45 % Saturated Hydraulic Conductivity*, Ks=1.0E-05 cm/sec	$\theta = \theta_r + \frac{\theta_s - \theta_r}{\left[1+\left(\frac{\psi}{a}\right)^n\right]^m}$ a = 6.13 kPa n = 1.08 θ_r = 0.08 cm3/cm3	$\theta = C(\psi)\frac{\theta_s}{\left\{ln\left[e+\left(\frac{\psi}{a}\right)^n\right]\right\}^m}$ a =162.62 kPa n = 0.79 m = 0.39

5.3 FLOW OF WATER THROUGH PAVEMENT LAYERS

During rainfall, the moisture content and the distribution of moisture within the pavement are not adequate enough (except in surface ponding from say flooding) to cause uniform saturated conditions in the base/subbase and subgrade layers (Ariza and Birgisson, 2002). Note that saturated conditions are used for design of permeable bases in the DRIP software, as discussed earlier. In most cases, unsaturated conditions exist, and under such conditions theories of saturated flow cannot be used to accurately predict the flow of water. For example, in a situation where there is rain after a long dry period, the pavement layers are under an unsaturated condition. This has been

TABLE 5.3
Index Properties and Estimated Parameters for Fine Grained Soils (With Kind Permission From Dr. Ghada S. Ellithy)

Soil Properties	Parameters for Van Genuchten, 1980 method	Parameters for Fredlund and Xing (1994) method
Void Ratio, e = 0.45 D_{10} = 0.1 mm D_{20} = 0.25 mm D_{30} = 0.42 mm D_{60} = 1.5 mm D_{90} = 5 mm Sand Content (< 2 mm) = 95 % Fines Content (< 0.075 mm) = 5 % Clay Fraction (< 0.002 mm) = 1 % Saturated Hydraulic Conductivity, Ks = 1.0E-03 cm/sec	$\theta = \theta_r + \frac{\theta_s - \theta_r}{\left[1 + \left(\frac{\psi}{a}\right)^n\right]^m}$ a = 0.99 kPa n = 1.58 θ_r = 0.06 cm³/cm³	$\theta = C(\psi)\frac{\theta_s}{\left\{ln\left[e + \left(\frac{\psi}{a}\right)^n\right]\right\}^m}$ a = 5.23 kPa n = 4.98 m = 0.81

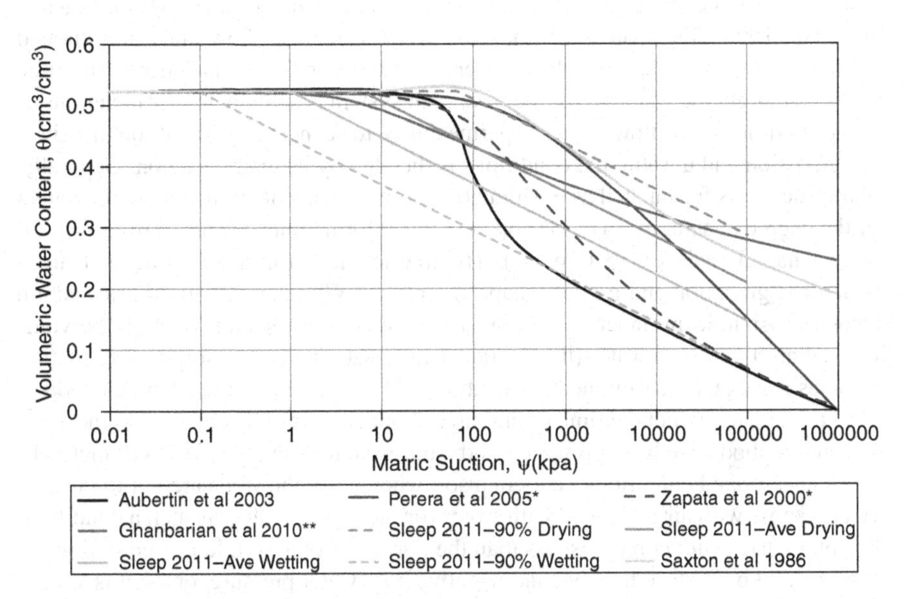

FIGURE 5.2 Examples of SWCC for fine grained soil using different methods.
Note: *Estimating Fredlund and Xing model parameters
**Estimating Van Genuchen model parameters (Courtesy, Dr. Ghada S. Ellithy, US Army Corps of Engineers. Used with permission.)

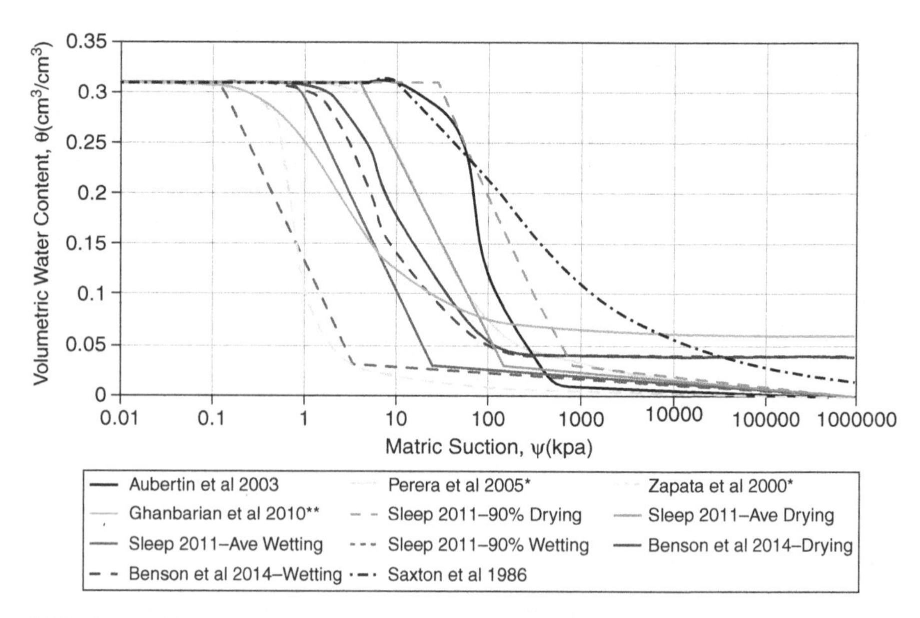

FIGURE 5.3 Examples of SWCC for coarse grained soil using different methods. (Courtesy, Dr. Ghada S. Ellithy, US Army Corps of Engineers. Used with permission.)

experimentally proven from field data from base courses in pavements (Roberson and Birgisson, 1998; Birgisson and Roberson, 2000). Generally, flow under unsaturated conditions is slower than the flow under saturated conditions, and hence the time required to drain the water out of the pavement system is actually longer under unsaturated conditions. The flow of water (permeability, to be more specific) through a given soil-gradation and unsaturated conditions is dictated by its matric suction. Generally, matric suction is formulated as a parameter that is dependent on the moisture content (or the degree of saturation) of the soil, and this information is obtained from the Soil Water Characteristic Curve (SWCC). The matric suction in a soil is dependent on its density, gradation and particle shape (Lamb and Whitman, 1969). Matric suction becomes insignificant under very high saturation conditions, and when all the voids in a soil are filled with water (that is, under fully saturated condition) matric suction becomes zero, and the saturated permeability (also referred to as hydraulic conductivity) is used, to predict the time of drainage. Applications of this method in the Time to Drain method have been presented in the discussion on the FHWA DRIP method.

An unsaturated condition occurs in a soil when all of the voids are not filled with water – some with air and some with water (or mostly with air). In such a condition, the volumetric water content is less than the porosity of the soil. In this condition the water is held by surface tension, and negative pore water pressure, or what is known as matric suction, is present. This suction pressure (or suction) in such a case controls the permeability of the soil. It is important to understand the difference between two phenomena here – one is the infiltration of the rainfall water into the soil, or the base/subbase, and the other is the flow of water inside the pavement system, once the water has infiltrated. The infiltration can happen from both above (joints, cracks and voids in the surface layer) or below (capillary rise, ground water table fluctuation).

The infiltration rate depends on the amount of water (from rainfall or flooding, such as depth of water), the porosity of the surface and the thickness of the layer.

Within the soil layer, if saturated conditions do not exist, then the permeability is reduced to a lower level. The relationship between the volumetric water content and the soil suction is expressed by the SWCC (Figure 5.4) and that between the suction and the permeability is expressed by the hydraulic conductivity curve. The shape of the SWCC is a function of its gradation, particle shape and packing. Two definitions are important – air entry value, that is the matric suction at which air starts entering the largest pores in the soil, and the residual water content, which is the water content at which large amounts of suction is required to remove the additional water from the initial saturated soil. The hysteresis effect can be seen in the difference of the SWCC curves in the wetting and drying (absorption and desorption phases) periods (Figure 5.5).

Note that both the volumetric water content and the degree of saturation are utilized along with matric suction-hydraulic conductivity to characterize a soil (Espinoza, 1993).

$$S_e = \frac{(S - S_r)}{(1 - S_r)} = \frac{(\theta - \theta_r)}{(\theta_s - \theta_r)} = \frac{(\theta - \theta_r)}{\Delta\theta} \quad (5.7)$$

S_e = effective degree of saturation
S is the degree of saturation,
S_r is the residual saturation corresponding to the value of θ_r
θ_r = the residual water content
θ_r is the residual water content
θ_s is the volumetric water content at saturation

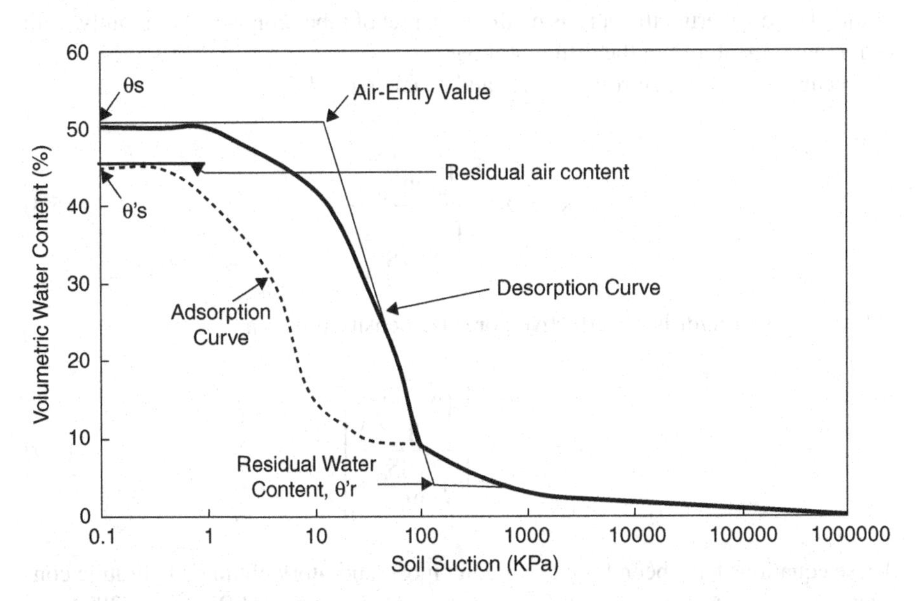

FIGURE 5.4 Typical SWCC. (Ariza and Birgisson, 2002. Courtesy: Minnesota Department of Transportation.)

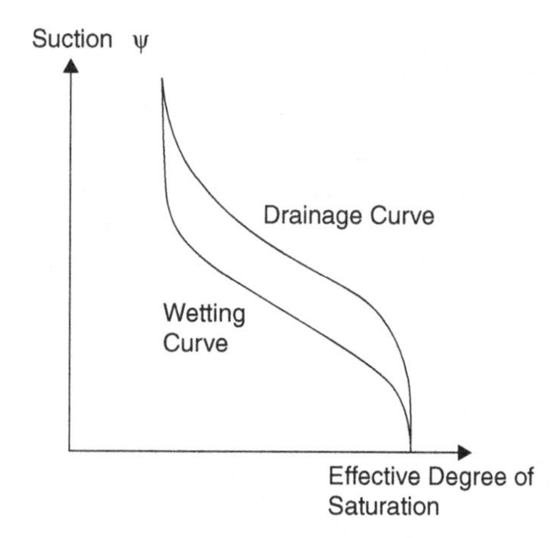

FIGURE 5.5 Schematic of hysteresis effect. (Ariza and Birgisson, 2002. Courtesy: Minnesota Department of Transportation.)

SWCC can be developed from data obtained from experiments. Hydraulic conductivity curves can also be generated from experimental data, but because of time and complexity, they are mostly developed from analytical models in combination with the SWCC data.

The different approaches for deriving the relationships between the degree of saturation or the volumetric water content and the matric suction are listed in Table 5.4. Table 5.5 compares the different models.

Relative permeability is defined as the ratio of unsaturated permeability to saturated permeability. This relative permeability has been related to a degree of saturation by considering the porous medium as a set of tubes connected randomly, with different permeability in the different tubes.

Examples are from Burdine (1953) and Mualem (1976).

$$k_r(S_e)=\frac{S_e^2\int_0^{S_e}\frac{dS_e}{\Psi^2(S_e)}}{\int_0^1\frac{dS_e}{\Psi^2(S_e)}} \tag{5.8}$$

Where, $S_e(r) = dS_e/dr$ is the effective pore size density function.

$$k_r(S_e)=S_e^2\left[\frac{\int_0^{S_e}\frac{dS_e}{\Psi(S_e)}}{\int_0^1\frac{dS_e}{\Psi(S_e)}}\right]^2 \tag{5.9}$$

These equations have been found to predict most laboratory obtained hydraulic conductivity curves reasonably well for a n value of 0.5 (Ariza and Birgisson, 2002).

TABLE 5.4
Different Models of SWCC (Adapted from Ariza and Birgisson, 2002; Courtesy: Minnesota Department of Transportation)

Brooks and Corey model (Brooks and Corey, 1964)

$$S_e = \left(\frac{\Psi}{PB}\right)^{\frac{-1}{v}} \text{ for } \Psi \geq PB$$

$$S_e = 1 \text{ for } \Psi < PB$$

Where,

Se is the effective degree of saturation
Ψ = the matric suction
PB = the bubbling pressure of the soil, which is the height of the capillary fringe
v = the pore size distribution index parameter, a measure of the soil grain uniformity

Bear model (Bear, 1972)

$$C_w = -\frac{\partial \theta}{\partial \psi}$$

Using

$$S_e = \frac{(S - S_r)}{(1 - S_r)} = \frac{(\theta - \theta_r)}{(\theta_S - S_r)} = \frac{(\theta - \theta_r)}{\Delta\theta}$$

Where,

S = the degree of saturation
Sr = the residual saturation corresponding to the value of θr
θ = the volumetric water content
θr = the residual water content
θs = the volumetric water content at saturation

And

$$S_e = \left(\frac{\Psi}{PB}\right)^{\frac{-1}{v}} \text{ for } \Psi \geq PB$$

$$S_e = 1 \text{ for } \Psi < PB$$

$$C_w = \frac{\Delta\theta \times S_e}{v\psi}$$

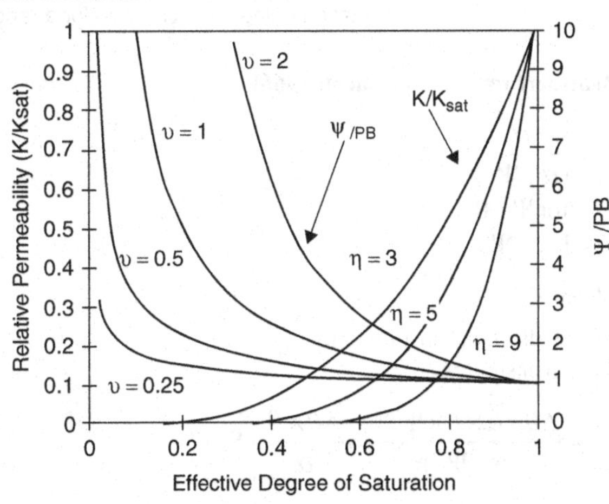

Influence of ν on Se – Ψ

(continued)

TABLE 5.4
Different Models of SWCC (Adapted from Ariza and Birgisson, 2002; Courtesy: Minnesota Department of Transportation) (Cont.)

Van Genuchten model (van Genuchten, 1980)

$$S_e = \frac{1}{\left(1+(\alpha\Psi)^{\beta}\right)^{\gamma}}$$

$$S_e = 1 \text{ for } \Psi < 0$$

Where,

β and γ = dimensionless coefficients, and γ = 1-1/β

α = a coefficient that has the dimension of the inverse of the piezometric head

$$C_w = \alpha \times \Delta\theta \times (\beta - 1) \times (\alpha \times \Psi)^{\beta-1} S_e^{1+1/\gamma}$$

C_w = water capacity

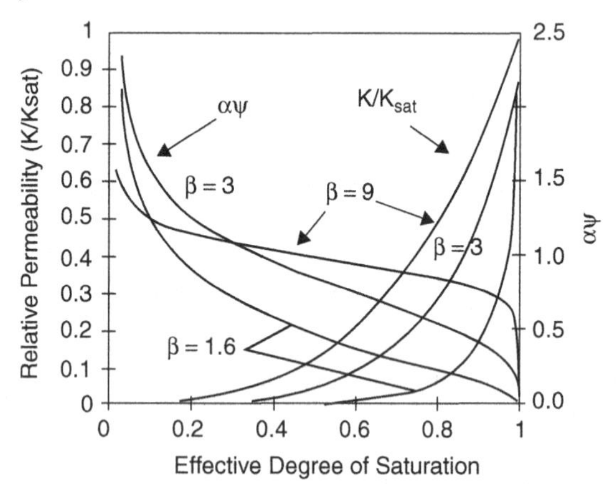

Variation of the matrix suction for a range of b values

Brutsaert model (Brutsaert, 1966)

$$S_e = \frac{\alpha}{(\alpha + \Psi^{\beta})} \quad \text{for } \Psi > 0$$

$$S_e = 1 \text{ for } \Psi < 0$$

Where,

α depends on the suction units

β is dimensionless

$$C_w = \frac{\Delta\theta \times \alpha \times \beta \times \Psi^{\beta-1}}{(\alpha + \Psi^{\beta})^2} = \frac{\Delta\theta \times \beta}{\alpha} \times S_e^2 \Psi^{\beta-1}$$

Where α and β are empirical coefficients.

TABLE 5.4
(Cont.)

Vauclin model (Vauclin et al, 1979)

$$S_e = \frac{\alpha}{(\alpha + \ln(\Psi)^{\beta})} \text{ for } \Psi > 1$$

$$S_e = 1 \text{ for } \Psi \leq 1\,\text{cm}$$

$$C_w = \frac{\Delta\theta \times \alpha \times \beta \times \ln(\Psi)^{\beta-1}}{\Psi(\alpha + \ln(\Psi)^{\beta})^2} = \frac{\Delta\theta \times \beta}{\alpha \times \Psi} \times S_e^2 \times (\ln\Psi)^{\beta-1} \text{ for } \Psi \geq \text{cm}$$

Where α and β are fitting coefficients.

Bear and Verruijt model (Bear and Verruijt, 1987)

This model presents the ink-bottle and the rain-drop effect to explain hysteresis (Figure 5.5), which causes the value of suction during the drainage to be greater than during wetting. These effects are due to the variation in the radii of the meniscus and that of the contact angle.

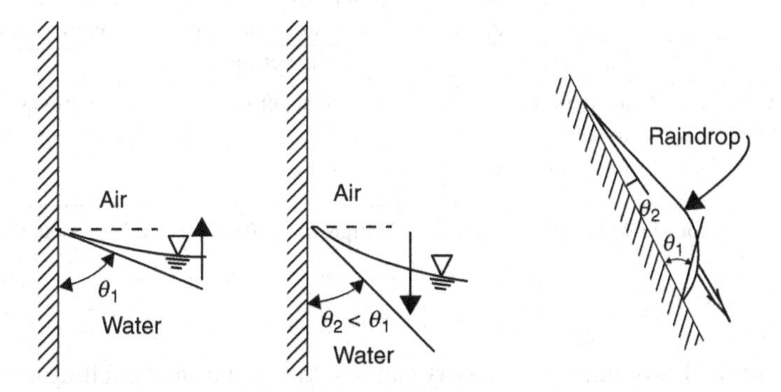

FIGURE 5.6 Ink bottle and raindrop effects. (Ariza and Birgisson, 2002. Courtesy: Minnesota Department of Transportation)

Mualem model (Mualem, 1973) proposes the porous medium as a system of pore domains with characteristic wetting and drying poor radii.

$$\theta = \iint f(\Psi_w, \Psi_d)\, d\Psi_w\, d\Psi_d$$

Where,

Ψ_w and Ψ_d are independent variables representing the suction on the main wetting and drying curves

Ψ_w always greater that or equal to Ψ_d

f is a bivariate probability distribution function

Source: Ariza and Birgisson, 2002. Courtesy, Minnesota Department of Transportation.

TABLE 5.5
Comparison of Different Models

Model	Main Features	Advantages	Disadvantages
Brooks and Corey	Relates the degree of saturation and the matric suction	Based on substantial experimental data	Empirical method
Brutsaert	Relates the degree of saturation and the matric suction	Sensitive to suction values close to saturation	Empirical coefficients
Mualem	Determines the volumetric water content as a function of the suction on the main wetting and drying curves	Based on rigorous statistical considerations and hysteresis effect included	Requires wetting and drying test data
Vauclin	Relates the degree of saturation and the matric suction	Describes laboratory measurements at small suction well	Empirical coefficients
Bear and Verruijt	Considers the hysteresis phenomenon	Hysteresis effect included	Requires wetting and drying data

Source: Adapted from Table 2.3, page 18, Ariza and Birgisson, 2002. Courtesy, Minnesota Department of Transportation.

The saturated hydraulic conductivity can be estimated from the falling or constant head permeameter tests. For soils with lower k, flexible or rigid wall permeameter or oedometers can be used, and higher pressure differences (with pump) may be needed. Unsaturated permeability can be estimated with the transient flow in soil columns, pressure plate outflow technique, osmotic permeameter and permeameter.

Hydraulic conductivity can also be estimated from models, as summarized in Table 5.6. Table 5.7 compares the different models (Ariza and Birgisson, 2002).

It should be noted that the presence of different layers (fine and coarse grained aggregates) has a significant effect on the drainage capacities of pavements. Because of higher suction and corresponding detrimental effect on drainage in fine grained soils, when fine grained soils are above coarse grained soils in a pavement, there is a reduction in flow through such a layered system. A capillary barrier tends to form between the fine and the coarse grained layers because of the relative difference in the drop in hydraulic conductivity with a change in pore suction levels in the two types of materials. The net resultant effect may cause most of the water to flow through the fine grained soil but remain at the intersection of the fine and the coarse grained soil (due to the barrier effect). Whether this happens or not depends on a number of factors, such as the sensitivity of the hydraulic conductivity to the change in suction and the relative location of the

TABLE 5.6
Different Hydraulic Conductivity Models

Gardner (1956) (Coefficients for fitting curves are obtained from experiments)

$$k(\psi)=\frac{k_s}{(1+A_k.\psi^{\beta}}$$

Where, k_s is the saturated hydraulic conductivity, ψ is the pore suction, A_k, β are empirical curve fitting coefficients

---

Brooks and Corey (1964)

$$k(\theta)=k_s\left(\frac{\theta-\theta_r}{\theta_s-\theta_r}\right)^n$$

where, k_s is the saturated hydraulic conductivity

θ is the volumetric water content

θ_s is the saturated volumetric water content,

θ_r is the residual volumetric water content,

$n = 3 + 2/\lambda$, and is based on the pore size index (λ). λ can be obtained from a regression function (Rawls et al, 1992) which is based on the percent sand, the percent clay, the soil porosity, and the cross product of these values.

Green and Corey (1971)

$$k(\theta)_i=\frac{k_s}{k_{sc}}\cdot\frac{30\gamma^2}{\rho g n}\cdot\frac{\varepsilon^P}{n^2}\sum_{j=1}^{m}\left[(2j+1-2i)\psi_j^{-2}\right]$$

for $i=1, 2,..., m$

where

$k(\theta)_i$ is the calculated conductivity for a specified water content,

θ is the volumetric water content,

i denotes the last water content class on the wet end, e.g. i = 1 identifies the pore class corresponding to the lowest water content for which the conductivity is calculated,

k_s/k_{sc} is a matching factor,

γ is the surface tension,

ρ is the density of water,

g is the gravitational constant (cm/sec2),

η is the viscosity of water (g/cm/sec-1),

ε is the soil porosity (cm3/cm3),

p is a parameter that accounts for interaction of pore classes,

n is the total number of pore classes between $\theta = 0$ and θs,

ψ_j is the pore pressure for a given class of water filled pores

(continued)

TABLE 5.6
Different Hydraulic Conductivity Models (Cont.)

van Genuchten (1980)

$$k(\theta) = k_s\left(\frac{\theta-\theta_r}{\theta_s-\theta_r}\right)^{\frac{1}{2}}\left[1-\left(1-\left(\frac{\theta-\theta_r}{\theta_s-\theta_r}\right)^{\frac{1}{m}}\right)^m\right]^2$$

where:

k_s is the saturated hydraulic conductivity,

θ is the volumetric water content,

θ_s is the saturated volumetric water content,

θ_r is the residual volumetric water content,

$m = \lambda / (1 + \lambda)$

λ = pore size index

TABLE 5.7
Comparison of Different Hydraulic Conductivity Models

Model	Main Features	Advantages	Disadvantages
Gardner	A two-parameter empirical model	Permeability is represented as a smooth function of suction instead of degree of saturation	Empirical curve-fitting coefficients must be acquired from experimental hydraulic conductivity data
Brooks and Corey	Allows the inclusion of the soil-water characteristic curve in a direct way	Based on substantial experimental data	Empirical equation, not very efficient at low suction values, relies on a difficult to obtain λ value
Green and Corey	Refined equation to model the hydraulic conductivity	Based on substantial experimental data	Use of many soil and fluid properties
Van Genuchten	Equation based on the volumetric water content	Based on substantial experimental hydraulic conductivity data at low suction values	Relies on a difficult to obtain λ value

Source: Adapted from Table 2.4, page 23, from Ariza and Birgisson, 2002. Courtesy, Minnesota Department of Transportation.

suction value with respect that at which the hydraulic conductivities of the coarse and the fine grained soils are similar. Due to these complexities, many authors have analyzed the problem of the flow of water through pavements through a simulation approach rather than analytical approach.

A number of researchers have developed approaches and framework for using unsaturated flow for the prediction of drainage time in pavements. These approaches use finite element (FE) or finite difference (FD) methods, and have been used to examine the effect of open graded bases, edgedrains, underdrains or combinations of the two, under different types of soils and materials. Ariza and Birgisson (2002) utilized actual water data from rainfall information and volumetric water content from layers (using time domain reflectometry, TDR, probes) in layers of experimental test sections in MN/ROAD experimental facility.

They used SEEP/W to model drainage in pavements using unsaturated and unsaturated-saturated transition flow. They also compared the time to drain results from this analyses to those obtained from the DRIP analysis. In the case of the SEEP/W analysis, Richard's equation (Richards, 1974) is used as a more convenient form of Darcy's Law, as follows:

$$\frac{\partial}{\partial x}\left(k_x\frac{\partial H}{\partial x}\right)+\frac{\partial}{\partial y}\left(k_y\frac{\partial H}{\partial y}\right)+Q\frac{\partial \Theta}{\partial t} \tag{5.10}$$

where
H = total head,
k_x = hydraulic conductivity in the x-direction,
k_y = hydraulic conductivity in the y-direction,
Q = the applied boundary flux,
Θ = volumetric water content,
t = time.

Test sections from the Mn/Road facility were modeled in this software, which consisted of three sections with approximately 4 inches of HMA over 12 inches of crushed granite (MNDOT Class 6 special gradation) base and R-12 sandy clayey silty soil subgrade. The HMA layers were considered as impervious layers and the unpaved shoulders were considered as the only sources of water. The SWCC and hydraulic conductivity curves (from which k_{sat} was estimated at zero suction) were estimated from suction plate measurements. In this analyses, the authors simulated a specific saturation level by placing the water level at a certain depth, which resulted in a suction in the base course (which resulted in effective saturation of 50%). In another case, they simulated the model by considering a 100 times more permeable material as a subbase underneath the base course layer. This resulted in a much faster drainage of water from the base layer, an approximate 10% drop in volumetric water content of the base course near the edge of the unpaved shoulder and the beginning of the paved surface.

The authors reported reasonable accuracy from the FEM models through comparisons with results predicted by analytical methods. They also modeled Mn/Road test sections exactly as they were with different materials properties, compared the model results with those obtained from the field data, and then back calculated the hydraulic properties to match the simulated results to the field data (volumetric moisture contents). The base and subgrade layers were represented by quadrilateral and triangular finite elements. An actual rain event that was recorded in the field was simulated in the model. The initial water table was placed at a depth so as to obtain the specific suction value that matched against the initial volumetric water content that was obtained from the field data. Zero heads were applied at the bottom corners of the model to initiate lateral and vertical drainage in the pavement.

The analysis consisted of two parts – a steady state analysis and a transient analysis. The steady state analysis was conducted to obtain the initial head condition of the system. In this analysis, the initial total head was set on the vertical sides of the subgrade to obtain water table elevation, which provided the suction values, which matched with the initial volumetric water contents (from the field data). At the end of this analysis the depth to the water table was established (from the top of the pavement). The water table found from this analysis as well as from the actual measurements underneath the paved part was found to be higher than that seen under the shoulder. Next the transient analysis was performed in which the rain event was simulated (as flux, m/s per m^2) and water was allowed to infiltrate into the pavement system.

The volumetric water content from the analysis was captured at the specific depths of the three locations of the base layer as were used in the actual test section in Mn/Road. The measured volumetric water contents were higher underneath the paved parts compared to the unpaved parts.

Now, since the model was setup initially to simulate the initial volumetric water contents in the layers, and the actual rainfall data were simulated, it was expected that the volumetric water content during the rainfall events, as captured from the field, will be exactly replicated in the simulation results. This was not found to be the case, and the authors postulated that the lab measured SWCC curves needed to be adjusted to yield the volumetric water content that would match the field data. They did however, note that part of the difference could be due to the limitation of the TDR technology that make them report variable moisture content under the same conditions for the same material. The adjustment or back calculation was done by changing the air entry value and changing the slope of the SWCC curves. A difference in air entry value between the lab and field soil can be expected due to a difference in the density of the material resulting from lab and field compaction. Next, taking these adjusted SWCC, the SEEP/W software was used to generate the hydraulic conductivity curves using the Green and Corey (1971) approach. Note that the authors also adjusted the percentage of rain water that was infiltrating into the pavement – initially they had assumed 100% to be going inside the surface of the pavement. Note that the model predicted volumetric water contents in the post rainfall event were found to be higher than the actual field data, and the authors had to do multiple adjustments for the soil subgrade permeability to finally match the predicted data and the field data. The authors also changed the percentage of infiltration for the different days, and the saturated volumetric moisture content to match the predicted data and the actual data – finally they commented that even though exact matching was not obtained, the trends were found to be similar. The authors noted that on the basis of the adjustments needed to match the predicted and the actual volumetric water content, it can be inferred that 30% of precipitation actually infiltrated the system, in spite of the presence of the HMA layer (4 inches) on the surface (which is often assumed to be an impervious surface).

The authors then proceeded to do a parametric study to evaluate the effects of various factors of the base and subgrade materials on the predicted volumetric water content in the base material. They reported the following findings:

1. A change in the initial slope of the base SWCC does not affect the results significantly.
2. Base Air entry value has a significant effect on the water content (a change from 4 to 5 kPa doubled the predicted maximum volumetric water content from 12.2–23.4%); however the results were insensitive to the subgrade air entry value.
3. A change in the k_{sat} of the base did not cause any significant change in the predicted water content, and a change of k_{sat} of the subgrade caused a slightly higher but still insignificant change to the volumetric water content

Of the different base gradations evaluated, the one with the highest saturated volumetric water content showed the highest volumetric water content (SWCC curve is higher), in spite of having the highest ksat value.

The authors therefore warn against solely relying on k_{sat} values for ensuring fast drainage. They reported that the Class 6 Special materials (MNDOT) performed the best in terms of drainage in which the volumetric water content had the highest difference from the saturated volumetric water content.

The predicted moisture content was found to be moderately sensitive to the infiltration rate. The predicted water content was found to be sensitive to the changes in the location of the initial water table; as it was changed from 0.76–0.96 m (6.2%), the predicted maximum water content changed from 9.7–7.7% (20.6%).

The authors also compared the relative effects of edgedrains and underdrains through modeling. An edgedrain consists of a backfilled trench with collector pipes, placed under the shoulder, with the collector pipe connected to transverse drains located at specific intervals along the pavement. However, geotextile (0.02 m) underdrains, which are placed below the base course, are provided with longitudinal pipes and this system is embedded inside the pavement system rather than being exposed as in an edgedrain. Underdrains are typically wrapped with geotextiles. Note that what the authors refer to here as underdrain is actually a geocomposite consisting of ribbed structure inside two geotextile layers which can serve the function of a permeable base course. In one case they assumed no edgedrain or collector pipe connected to this geocomposite drain and in another case they considered it with the 0.1 m diameter collector pipes on both sides of the pavement. To simulate flow in the collector pipes, a pressure head of 0 m was set around the pipe circumference as a boundary condition. In a third case the authors assumed edgedrains only located underneath the base materials, running parallel to the pavement at the end of the paved area. Again a pressure head of 0 was applied around the pipe circumference as boundary condition to simulate flow. A fourth case was in which the underdrain (geocomposite) was connected to an edgedrain (longitudinal drains system). It was observed from the analysis that the system in which the edgedrain was connected to a collector or edgedrain and was kept underneath the well-draining Class 6 special material was the most effective for drainage. There was no significant difference between the presence of conventional edgedrains of collector pipes. The authors postulated that the method is effective by shortening of the drainage path and affecting the distribution of suction in the base material.

Finally, the authors stress the importance of the following:

1. Model drainage through unsaturated flow theory.
2. Accurately estimate water table depth and variation of moisture content in existing pavements both along the vertical and the horizontal axes.
3. Utilize horizontal arrays of TDRs along the layer interfaces to understand the unsaturated flow through pavements.

The two critical components of a positive pavement drainage system are as follows.

1. Open graded base with coarse gradation, protected by filters from clogging (ingress by piping) by fine particles from the lower (subgrade layers), and connected to an underdrain system with a longitudinal channel and periodic collector pipes. If the structural strength of the base course is found to be low, it can be stabilized with asphalt or Portland cement which will result in reduction in permeability. Note that some states have reported no improvement in drainage from this approach, while others have reported positive results, and in many cases clogging has been found. Also, the provision of this layer can result in an increased thickness of the pavement section because of the thickness of the permeable base course that is required for the flow of the water and also because of the need for compensating the low structural strength of such a layer.
2. The edgedrains in the underdrain system can be built with materials such as concrete or plastic and can be perforated. Generally, they are backfilled with coarse aggregates of appropriate size using proper compaction (to avoid settlement over the drain) and wrapped with geotextiles to allow ingress of water but prevent clogging by fines. Such a drainage channels are placed at a longitudinal slope outside of the wheel path and the trackline of the paver.

5.4 ANALYTICAL MODEL FOR PREDICTING THE FLOW OF WATER IN PAVEMENTS

Recently, Kalore et al. (2019) have presented an analytical model for estimating the time to drain considering the unsaturated characteristics of pavement base material, and demonstrated that the model performs as good as the finite element analysis. (Note that the Barber and Sawyer method (1952) and Casagrande and Shannon method (1952) that are utilized in the DRIP program were developed on the assumption of full saturation in the permeable base layer.) The key model parameters include the material property and geometric-section properties of the drainage layer in the pavement subsurface. Material properties include drainage characteristics: SWCC and hydraulic function whereas the geometric-section properties include the flow length and flow gradient. The model is capable of simulating the precipitation-infiltration-drainage process for a given rainfall event and can quantify the temporal moisture content distribution. The temporal moisture content distribution is an important input factor in quantifying the pavement deterioration and the resilient modulus for the structural design of pavement layers. Another important result of the developed model is the independence of the layer thickness with the time to drainage (as reported in the literature) and states that it is reflected in terms of moisture retaining capacity of the drainage layer given that the permeability is not affected by layer thickness.

5.4.1 Model Description

The average variation of matric suction along the flow length is assumed as a generalized parabola and an analytical equation for estimating the time-to-drainage is derived in terms of geometric-section properties, drainage characteristics of the base material and a matric suction distribution parabola parameter. The assumption of the matric suction distribution along the flow length, on which the mathematical analysis is based is illustrated in Figure 5.7. The initial conditions are maintained the same as that of the FHWA model (Casagrande and Shannon, 1952) so as to compare the results and to evaluate the quality of the drainage layer in terms of time-to-drain. The bottom of the layer is assumed as an impermeable boundary; theoretically, it can be justified upon considering the relative differences between the saturated permeabilities of open-graded aggregate and fine subgrade, though in reality it is mainly influenced by the hydraulic behavior of the subgrade soil. Therefore, it is reasonable to assume the bottom of the drainage layer as impermeable and major moisture flow will occur in the horizontal direction to the edgedrains. The horizontal flow is also influenced by the longitudinal and cross-slope of the pavement section. Therefore, the resultant flow length is defined as the longest distance traveled by the moisture to reach the edgedrain and flow gradient is the gradient along the flow length. The flow length is estimated by using equation 5.13 and flow gradient by using equation 5.14. This reduces the 3-dimensional system to 2-dimension for simplified analysis.

$$L_R = W'\sqrt{1+\left[S_y / S_x\right]^2} \tag{5.11}$$

$$S_R = \sqrt{S_y^{\,2} + S_x^{\,2}} \tag{5.12}$$

Where,

L_R = flow length
S_R = flow gradient
W' = width of pavement section
S_y = longitudinal slope
S_x = transverse slope

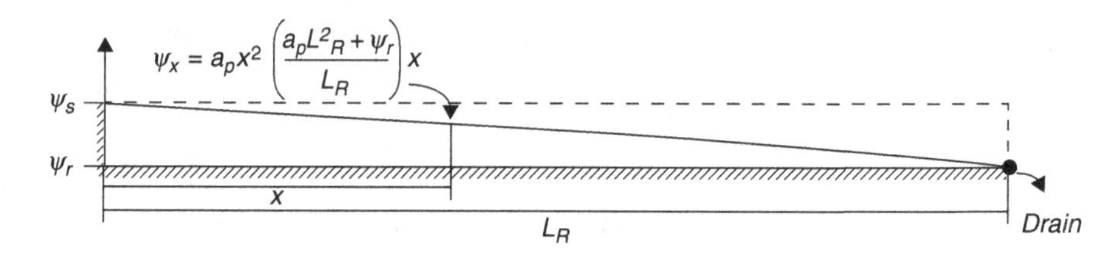

FIGURE 5.7 Drainage layer section: Developed Model.

From Figure 5.7, the matric suction (ψ_x) at any distance x along the flow length is given as,

$$\psi_x = a_p x^2 - \left(\frac{a_p L_R^2 + \psi_r}{L_R}\right)x \tag{5.13}$$

Where,

ψ_x = the residual matric suction
L_R = the flow length
a_p = the parabola parameter

Differentiating equation 5.15 gives:

$$\frac{d\psi_x}{dx} = 2a_p x - \left(\frac{a_p L_R^2 + \psi_r}{L_R}\right) \tag{5.14}$$

The rate of moisture content along the flow length can be also be written as equation 5.17:

$$\frac{d\psi}{dx} = \frac{dt}{dx}\frac{d\psi}{dt} = \frac{1}{dx/dt}\frac{d\psi}{dt} \tag{5.15}$$

Where, dx/dt is the permeability $k(\psi)$, which is a function of matric suction

Equating above equations,

$$\frac{d\psi}{dt} = k(\psi)\left(2a_p x - \left(\frac{a_p L_R^2 + \psi_r}{L_R}\right)\right) \tag{5.16}$$

The above equation can be written as,

$$dt = \frac{d\psi}{k(\psi)\left(2a_p x - \left(\frac{a_p L_R^2 + \psi_r}{L_R}\right)\right)} \tag{5.17}$$

Integrating both sides from full saturation to $U\%$ of the degree of drainage.

$$\int_0^{t_U} dt = \int_0^{\psi_U} \frac{d\psi}{k(\psi)\left(2a_p x - \left(\frac{a_p L_R^2 + \psi_r}{L_R}\right)\right)} \tag{5.18}$$

Simplifying the above equation and substituting x in terms of suction gives equation 5.20:

$$t_U = \frac{L_R}{k_s} \int_0^{\psi_U} \frac{d\psi}{k_r(\psi)\sqrt{(a_p L_R^2 + \psi_r)^2 + 4a_p L_R^2 \psi_U}} \tag{5.19}$$

Substituting equation 5.8 into equation 5.18:

$$t_U = \frac{L_R}{k_s} \int_0^{\psi_U} \frac{d\psi}{\left\{\frac{1}{\left[\left\{\ln\left[e + \left(\psi_U / a_f\right)^{n_f}\right]\right\}^{m_f}\right]}\right\}^{q_f} \sqrt{(a_p L_R^2 + \psi_r)^2 + 4a_p L_R^2 \psi_U}} \tag{5.20}$$

Equation 5.22 gives the time to drainage $U\%$ drainable moisture from full saturation for a given geometric section properties, base layer material drainage characteristics and parabola parameter (a_p).

5.4.2 Model Calibration

The a_p parameter can be calibrated based on the results obtained from the experimental or mechanistic approach. It is calibrated based on optimization such that the mean square error between the results obtained from the developed model and results obtained from finite element SEEP/W (2001) model is minimized. In FE analysis, the seepage flow is analyzed such that the difference between the inflow and outflow through a finite volume in a finite time is equal to change in moisture content. In this method the required inputs are layer domain, mesh characteristics and the unsaturated characteristics of the material. Time to drainage in SEEP/W is defined by setting up a parent analysis of steady state condition as full saturation without any drainage boundary condition for considered geometry. A clone transient analysis is setup under the parent analysis with similar conditions and with an additional point drainage boundary condition at a bottom corner so as to represent the edgedrain. Figure 5.8 shows the SEEP/W finite element mesh representation of the drainage layer section along the flow length for similar boundary conditions illustrated in Figure 5.7. The domain is discretized into 2-dimensional rectangular and triangular elements with a global element size of approximately 0.1 meter which results in the mass balance error of less than 1%.

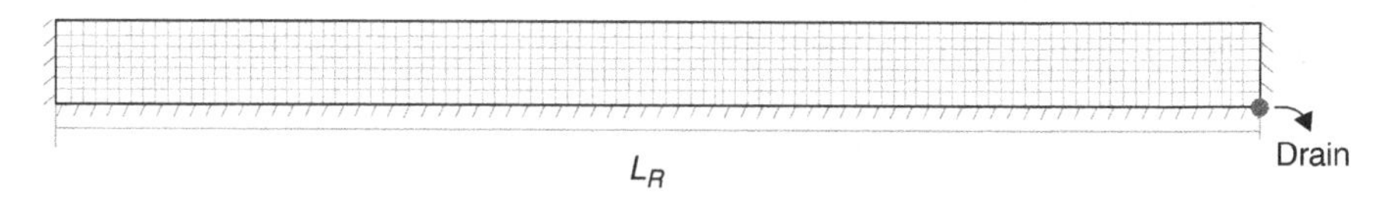

FIGURE 5.8 Drainage layer section: Finite Element Model: SEEP/W.

5.4.3 Model Application: Rainfall Simulation

The model helps in understanding the performance of the drainage layer for a given amount of inflow. The major parameters of precipitation specifically, its intensity and duration, govern the movement of moisture in pavement subsurface. Here, the performance is reflected in terms of the level of saturation produced by a given inflow. The lower the degree of saturation the lower will be the rate of deterioration of the pavement material and thus a lower life-cycle cost. The performance of the system can be improved by providing a drainage layer with sufficient drainage characteristics and section-geometry so that the system should have a lower degree of saturation even in the extreme inflow events through the design life of the pavement.

The degree of saturation produced by a rainfall event and other inflows can be estimated by simulating the process of infiltration and drainage simultaneously. The process of draining starts as soon as the moisture content of the drainage layer exceeds the residual moisture content. The produced moisture content by an inflow event can be estimated using the infiltration model. According to the Green-Ampt (1911) infiltration model, the amount and rate of infiltration depend on the inflow intensity and the saturated permeability of the pavement layer. The subsurface drainage system will be ineffective compared to the surface drainage system when the inflow intensity is more than the saturated permeability of the system, and this will cause a higher rate of deterioration of the material because of the undrainable infiltrated moisture in the system. In case of drainable bases with substantial permeability, the inflow intensity will always be less than the saturated permeability and the system will be unsaturated with a lower degree of saturation. This phenomenon is proved by simulating the infiltration and drainage simultaneously by considering the small-time step with the developed model as illustrated in the algorithm shown in Figure 5.9. The simulation will result in a variation of volumetric moisture content with time in a drainage layer. As the resilient modulus (M_r) is a function of moisture content, the simulation results are useful in quantification of variation of the modulus of the base layer in the pavement with time, so that pavement agencies can make decisions regarding the traffic flow considering the structural condition of the pavement during and after extreme precipitation (Mallick et al., 2017).

REFERENCES

Ariza, P. and Birgisson, B. 2002. *Evaluation of Water Flow through Pavement Systems*. University of Florida.

Aubertin, M., Mbonimpa, M., Bussiere, B. and Chapuis, R.P. 2003. A Physically-based model to predict the water retention curve from basic geotechnical properties. *Canadian Geotechnical Journal*, 40 (6): 1104–1122.

Bear, J. 1972. *Dynamics of Porous Media*. Dover Publication Inc. NY.

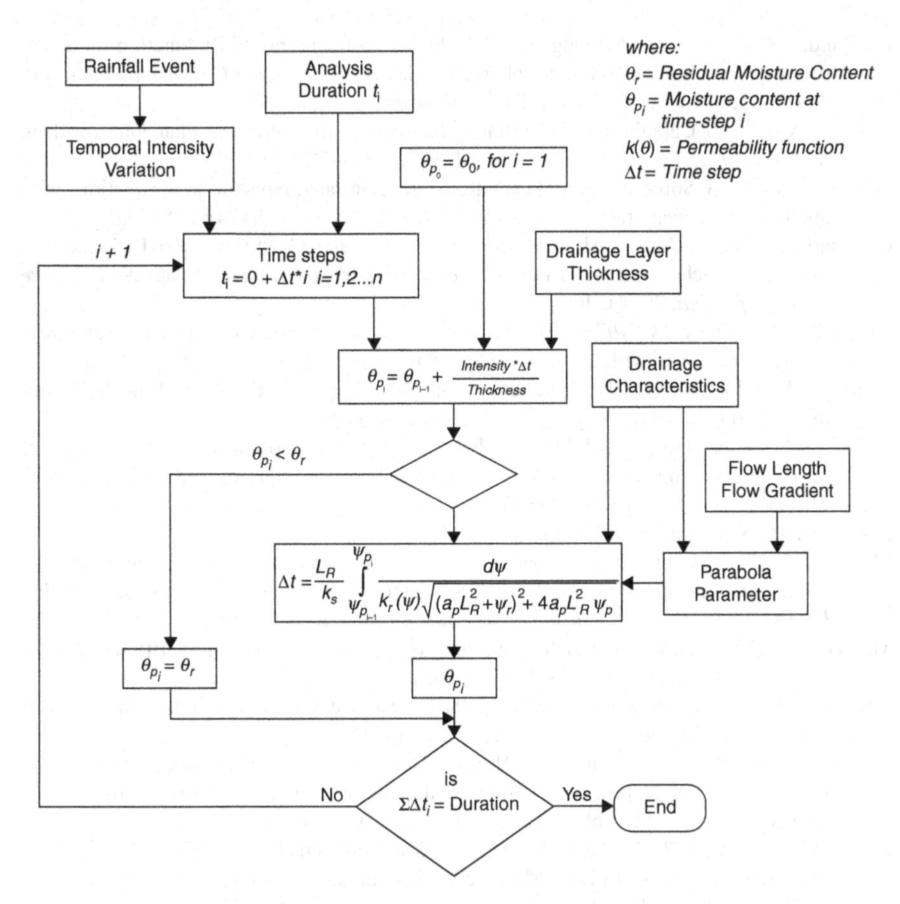

FIGURE 5.9 Simulation algorithm for drainage process (Kalore et al., 2019).

Bear, J. and Verrujt, A.M. 1987. *Modeling Groundwater Flow and Pollution*. D. Reidel, Dordrecht, The Netherlands.

Benson, C., Chiang, I., Chalermyanont, T. and Sawangsuriya, A. 2014. Estimating van Genuchten parameters α and n for clean sands from particle size distribution data. In *Soil Behavior Fundamentals to Innovations in Geotechnical Engineering*. 410–427.

Birgisson, B. and Roberson, R. 2000. *Drainage of Pavement Base Material: Design and Construction Issues*. Transportation Research Record 938, Transportation Research Board, National Research Council. Washington, DC.

Brooks, R.H. and Corey, A.T. 1964. Hydraulic properties of porous media. *Hydrol.* Paper 3, 37pp.: Colorado State University. Fort Collins, Colorado.

Brutsaert, W. 1966. Probability Laws for Pore-Size distribution. Soil Sci., 101, pp. 85–92.

Burdine, N.T. 1953. Relative permeability calculations from pore size distribution data. *Trans. AIME*, 198: pp. 71–78.

Casagrande, A. and Shannon W.L. 1952. Base course drainage for airport pavements. *Proceedings of the American Society of Civil Engineers*, 77: 792–814.

Espinoza, R.D., Bourdeau, P.L. and White, T.D. 1993. *Pavement Drainage and Pavement Shoulder Joint Evaluation*. Numerical Modeling of Infiltration and Drainage in Pavements. Final Report. Purdue University, west Lafayette, IN.

Fredlund, D.G. and Xing, A. 1994. Equations for the soil-water characteristic curve. *Canadian Geotechnical Journal*, 31: 521–532.

Fredlund, D.G., Xing, A. and Huang, S. 1994. Predicting the permeability function for unsaturated soils using the soil-water characteristic curve. *Canadian Geotechnical Journal*, 31 (4): 533–546. https://doi.org/10.1139/t94-062.

Gardner, W.R. 1956. Calculation of Capillary Conductivity from Pressure Plate Outflow data. *Soil Science Society of America Journal*, 3 (20): 317–320.

Gardner, W.R. 1958. Some steady state solutions of unsaturated moisture flow equations with applications to evaporation from a water table. *Soil Science*, 85 (4): 228–232.

Ghanbarian, A.B., Liaghat, A., Huang, G.H. and Van Genuchten, M. Th. 2010. Estimation of the Van Genuchen soil water retention properties from soil textural data. *Soil Science Society of China*, 20 (4): 465.

Green, R.E. and Corey, J.C. 1971. Calculation of hydraulic conductivity: a further evalua-tion of some predictive methods. *Soil Science Society of America Proceedings*, 35: 3–8.

Green W.H. and Ampt, G. 1911. Studies of soil physics, part I –the flow of air and water through soils. *Journal of Agricultural Science*, 4: 1–24.

Kalore, S.A., Babu, G.L.S. and Mallick, R.B. 2019. Design approach for drainage layer in pavement subsurface drainage system considering unsaturated characteristics. *Transportation Geotechnics*, 18: 57–71.

Lambe, R. and Whitman, R. 1969 *Soil Mechanics*. New York: Wiley.

Mallick, R.B., Tao, M., Daniel, J.S., Jacobs, J. and Veeraragavan, A. 2017. A combined model framework for asphalt pavement condition determination after flooding. *Transportation Research Record (TRR): Journal of the Transportation Research Board*, 2639 (1): 64–72.

Mualem, Y. 1973. Modified approach to capillary hysteresis based on a similarity hypothesis. *Water Resources Research*, 9 (5), 1324–1331.

Mualem, Y. 1976. A new model for predicting the hydraulic conductivity of unsaturated porous media. *Water Resources Research*, 12 (3): 503–522.

Perera, Y.Y., Zapata, C.E., Houston, W.N. and Houston, S.L. 2005. Prediction of the soil-water characteristic curve based on grain-size-distribution and index properties. ASCE Geotechnical Special Publication. 130, Reston, VA: 49–60.

Rawls, W.J., Ahuja, L.R., Brakensiek, D.L. and Shirmohammadi, A. 1992. *Infiltration and soil water movement.* In D.R. Maidment (ed.) Handbook of Hydrology (pp. 5.1–5.5). McGraw-Hill, Inc., New York.

Richards, B.G. 1974. *Behavior of Unsaturated Soils. In Soil Mechanics-New Horizons*, Chapter 4. New York: American Elsevier Publishing Company Inc.

Roberson, R. and Birgisson, B. 1998. "Evaluation of Water Flow Through Pavement Systems." *Proceedings. International Symposium on Subdrainage in Roadway Pavements and Subgrades*, pp. 295–302.

Saxton, K.E., Rauls, W.J., Romberger, J.S. and Papendick, R.I. 1986. Estimating generalized soil-water characteristics from texture. *Soil Science Society of America Journal*, 50 (4): 1031–1036.

SEEP/W. 2001. Users Manual, Version 4.24. GEO-SLOPE International Ltd. Calgary.

Sleep, M.D. 2011. Analysis of transient seepage through levees. PhD diss. Virginia Polytechnic Institute and State University, Blacksburg, VA.

van Genuchten, M. Th. 1980. A closed-form equation for predicting the hydraulic conductivity of unsaturated soils. *Soil Science of America*, 44: 892–898.

Vauclin, M., Khanji, D. and Vachaud, G. 1979. Experimental and numerical study of a transient two dimensional unsaturated-saturated water table recharge problem. *Water Resources Research*, 15 (5), 1089–1101.

Zapata, C.E., Houston, W.N., Houston, S.L. and Walsh, K.D. 2000. Soil-water characteristic curve variability. ASCE Geotechnical Special Publication 99. Denver, CO: 84–124.

6 The Impact of Water in Pavements

6.1 THE IMPACT OF WATER ON PAVEMENT DISTRESS AND CONDITION

The function of subsurface drainage is to control and/or drain subsurface water or moisture away from the pavement system especially the subgrade and subbase. Water may come through the joints or cracks of the pavement surface courses and from a high water table. Moisture can come through a capillary rise below the pavement. Subsurface drainage largely comprises of a permeable subbase drainage course within the pavement and drainage channels such as longitudinal and/or transverse pavement drains. Inadequate subsurface drainage causes premature failure of both concrete and asphalt pavements.

In the case of concrete pavements, if the underlying subgrade or subbase is wet, it loses strength (especially if it is clayey in nature) and settles under traffic loads creating a void space below the concrete slab. Water fills in the void space, and under traffic mud pumping occurs through joints or cracks in the concrete pavements. These void spaces continuously become larger due to pore pressure created by traffic loads. Soon, due to a lack of support this results in the faulting and cracking of concrete slabs especially at the corners. Although mud jacking can be done to eliminate or minimize these void spaces, it is a rather tedious process. Concrete pavements deteriorated by such a phenomenon soon require major rehabilitation. Use of asphalt treated permeable base (ATPB) over pumping concrete pavement prior to major asphalt overlay has been tried successfully by the Pennsylvania Department of Transportation on Interstate 80 (Figure 6.1). The ATPB at least relieves the localized force of pumping water and accommodates it without causing any harm to the overlying asphalt courses. If deemed proper, ATPB can be placed after cracking/seating or rubblizing concrete slabs. Obviously, free-draining soil beneath the concrete pavement will not provide the necessary water for pumping.

In case of asphalt pavements, inadequate subsurface drainage provides water or moisture, or moisture vapor, which is the necessary ingredient for inducing stripping in the asphalt mix. Stripping can be described as a loss of bond between the asphalt binder and the aggregate surface. Once the asphalt binder film peels off the aggregate surface, the asphalt mix loses cohesion and starts to disintegrate usually causing potholes. There are two primary mechanisms of stripping; both require water,

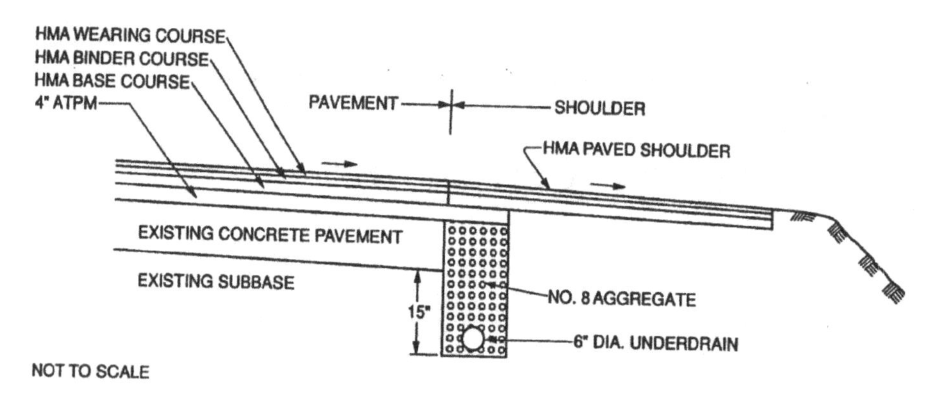

FIGURE 6.1 Use of ATPB over deteriorated, pumping concrete pavement.

moisture or moisture vapor. First, many aggregates are hydrophilic (water loving). When asphalt mix containing such aggregate comes in contact with water, asphalt film is displaced by water and stripping results. Second, if the compacted asphalt mix is saturated with water, pore pressure can be created under traffic physically removing (scouring) the asphalt binder from the aggregate surface (Kandhal, 1994).

Kandhal et al. (1989), and Kandhal and Rickards (2001), have reported case histories where the stripping was not a general phenomenon occurring on the entire project but rather a localized phenomenon in areas of the project over-saturated with water and/or water vapor due to inadequate subsurface drainage conditions. These case histories are described later. First, the background is presented below.

Water can enter the asphalt pavement layers in different ways. It can enter as run-off through the road surface, primarily through surface cracks. It can enter from the sides and bottom as seepage from ditches and a high water table in the cut areas. The most common water movement is upward by capillarity under a pavement. Above the capillary fringe water moves as a vapor. Many subbases or subgrades in the existing highway system lack the desired permeability, and, therefore, are saturated with the capillary moisture. The construction of multilane highways (or widening) to greater widths, gentler slopes and milder curves in all kinds of terrain has compounded the subsurface drainage problem. Doubling the road width, for example, makes drainage about four times as difficult as before (Cedergren and Lovering, 1968). Quite often, a four-lane highway is rehabilitated by paving the median and shoulders with asphalt resulting in a fully paved width of 22–24 m (72–78 feet) which is equivalent to a six-lane highway, without any increase in the subsurface drainage capability.

Extensive research has been conducted on the mechanism of asphalt stripping at the University of Idaho (Lottman, 1971). It has been reported that:

> Air voids in asphalt concrete may become saturated with water even from vapor condensation due to water in the subgrade or subbase. A temperature rise after this saturation can cause expansion of the water trapped in the mixture voids resulting in significant void pressure when the voids are saturated. It was found that void water pressure may develop to 20 psi under differential thermal expansion of the compacted asphalt mixture and could exceed the

adhesive strength of the binder aggregate surface. If asphalt concrete is permeable, water could flow out of the void spaces under the pressure developed by the temperature rise and, in time, relieve the pressure developed. If not, then the tensile stress resulting from the pressure may break adhesive bonds and the water could flow around the aggregates causing stripping. The stripping damage due to void water pressure and external cyclic stress (by traffic) mechanism is internal in the specimens, the exterior sides of the specimens do not show stripping damage unless opened up for visual examination.

The pore pressure from stresses induced by traffic causes the failure of the binder-aggregate bond. Initially, the traffic stresses may further compact the mixture and trap or greatly reduce the internal water drainage. Therefore, the internal water is in frequent motion (cyclic) and considerable pore pressure is built up under the traffic action (Majidzadeh and Brovold, 1966).

The internal water pressure required for causing a compacted asphalt mixture to have adhesive or interfacial tension failure (stripping) is inversely proportional to the diameter of the pores (Hallberg, 1950). Asphalt binder course mixtures generally strip more than the wearing course mixtures possibly due to large diameter pores in the binder course. Moreover, the wearing course is exposed to repeated high temperature drying periods when the pavement heals. The asphalt binder films which debond from the aggregate attach themselves again and the mix regains its strength and water resistance. The humid periods are longer in the underlying binder course and, therefore, the self-healing forces during warm periods have much less influence.

If there is insufficient drainage, water may saturate the base and rise through the pavement. The water that enters the structural section from below can create many drainage problems and result in deteriorated pavements. Apparently, the deterioration is caused by premature stripping in many cases (Lovering and Cedergren, 1962).

Telltale signs of water damage to asphalt overlays (over concrete pavements) have been described by Kandhal and Rickards (2001). They observed wet spots on the asphalt overlay surface scattered throughout the project. Usually at these wet spots water oozed out during hot afternoons. Some of the wet spots contained fines suspended in the water, which were tracked on the pavement by the traffic and appeared as white spots. These fines came from the aggregate surface exposed after stripping. Most white spots turned into fatty areas (resulting from asphalt binder stripping and migrating to the surface) which usually preceded the formation of potholes. Figure 6.2 shows all three stages: white spots, fatty areas and potholes on a four-lane highway. Figure 6.3 shows severely stripped aggregate particles in a pothole.

Small and large asphalt blisters have also been observed on the road surface due to entrapped moisture. A very severe case of blistering from moisture vapor pressure at Emporia Airport, Virginia has been described by Acott and Crawford (1987). However, blisters can also occur without any asphaltic globules at the surface.

Usually, stripping in a four-lane highway facility occurs first in the slow traffic lane because it carries more and heavier traffic compared to the passing lane. Typically, stripping starts at the bottom of the asphalt layer and progresses upwards. It is evident from the preceding discussion that inadequate subsurface drainage is one of the

FIGURE 6.2 Three stages of stripping on Interstate 40, Oklahoma, USA.

primary factors inducing premature stripping in asphalt pavements resulting in its disintegration. Very few case histories on the impact of inadequate subsurface drainage are found in the literature. Only some case histories have been discussed in detail (Kandhal et al., 1989; Kandhal and Rickards, 2001). These are described briefly here.

6.1.1 East-West Pennsylvania Turnpike, USA

A section of the East-West Pennsylvania Turnpike between Philadelphia and Pittsburgh was rehabilitated in 1977. This section received 100 mm (4 inch) asphalt overlay on the mainline and its 3 m (10 feet) wide median was also paved for the first time with a 75 mm (3 inch) asphalt binder and wearing course. The shoulders were also paved with dense graded asphalt mix. The work also included the installation of new pipe in the median. However, the new subbase above the pipe was almost impermeable. Figures 6.4 and 6.5 show typical median and cut sections of the East-West Pennsylvania Turnpike, respectively.

Stripping was observed in this pavement during the summer of 1978 when small potholes started to develop mainly in the inside wheel track of the slow traffic lane (Figure 6.6). The existing pavement had a pavement edgedrain as shown in Figure 6.5. However, it did not seem to be effective to drain the entire outside slow lane on the right. It was observed from extensive trenching and sampling that water and/or water vapor was getting into the pavement structural system from underneath, primarily through the longitudinal and transverse joints, cracks in the concrete

FIGURE 6.3 Close up of stripped asphalt mix in pothole on Interstate 40, Oklahoma, USA.

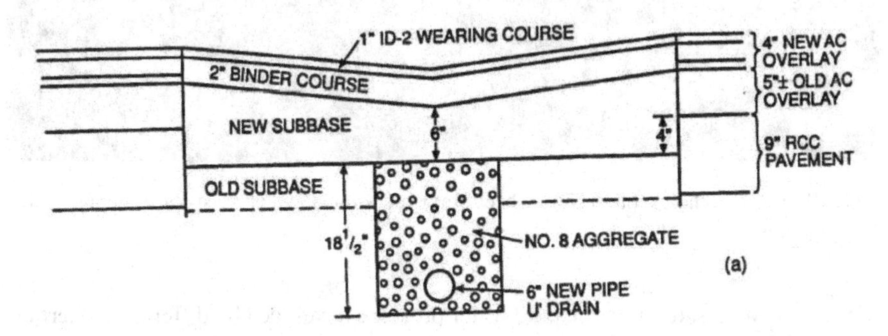

FIGURE 6.4 Paved median of East-West Pennsylvania Turnpike, USA.

pavement and the disintegrated concrete itself at some places. There was also evidence that moisture was being drawn from the subbase under the paved median into the asphalt overlay layers probably in the form of water vapor during the heat of the day (Figure 6.4). Water vapor which accumulated in the pavement layers during the day condensed during the night until the asphalt pavement layers become saturated

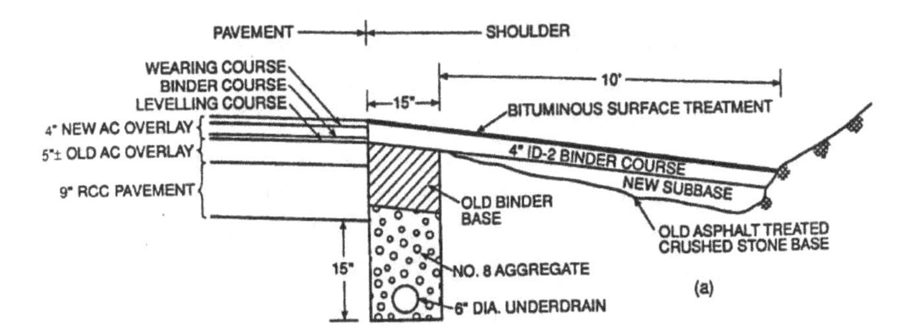

FIGURE 6.5 Typical cut section of East-West Pennsylvania Turnpike, USA.

FIGURE 6.6 Potholes on inside wheel track of slow lane of East-West Pennsylvania Turnpike, USA.

with water. With saturation the pore water pressure developed by differential thermal expansion and cyclic stresses from the traffic ruptured the asphalt-aggregate bond causing stripping.

Prior to major rehabilitation of this four-lane highway, subsurface moisture could partly escape through: longitudinal and transverse cracks of the pavement; uncovered median; and shoulders. This can be termed as "drainage by evaporation" (Figure 6.7). However, after the pavement, median and shoulders were all covered with relatively impervious asphalt course, subsurface moisture was entrapped under a 24 m (78 feet) wide black highway strip. This black surface would get heated up during the day and would draw moisture/moisture vapor upwards. Even dense graded asphalt mixes can

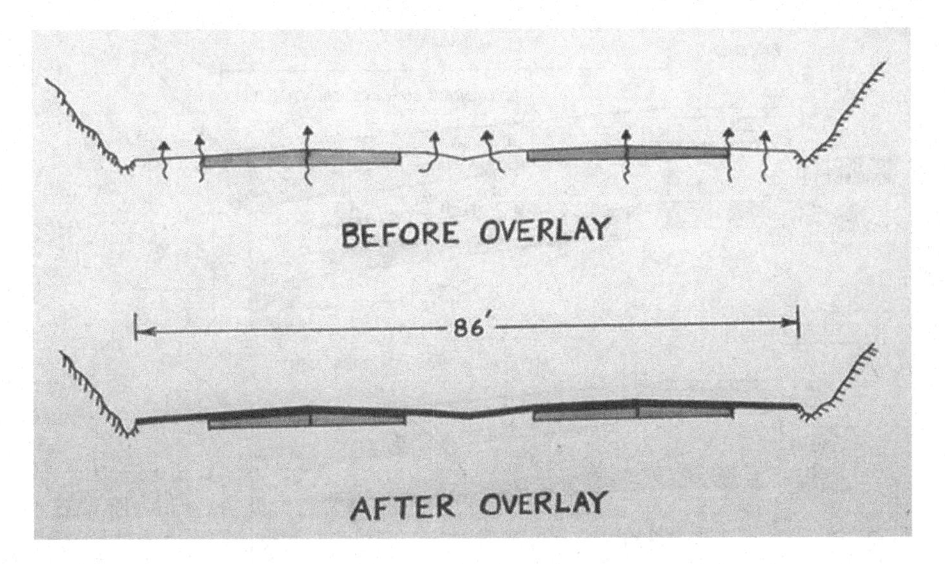

FIGURE 6.7 Drainage by evaporation prevented by 86 feet wide asphalt paving.

be somewhat permeable to moisture vapor, which causes stripping and subsequent disintegration.

6.1.2 North-East Extension of Pennsylvania Turnpike, USA

A section of the four-lane North-East Pennsylvania Turnpike consisted of 254 mm (10 inch) thick concrete pavement which was overlaid for the first time with 100 mm (4 inches) of asphalt overlay in 1977 and 1978. This section has a 1.2 m (4 feet) wide raised concrete median divider unlike the East-West Turnpike which had 3 m (10 feet) wide depressed median. The work also included providing 150 mm (6 inch) underdrain (U-drain) in the cut sections (Figure 6.8). The shoulders were also paved with dense graded asphalt course. All pavement lanes (both slow and passing lanes) in the cut areas started to exhibit white spots on the surface during summer of 1978. These white spots were formed by salt laden water oozing out in the afternoon on hot days. Although no potholes had developed in this section, there was concern that a distress pattern similar to that experienced on the East-West Turnpike may develop. Therefore, similar investigations were carried out. Extensive excavations indicated that the asphalt layers had substantial in-situ moisture but those near the concrete median were much wetter than those in the outside slow lane (right). Evidently, the new underdrain installed at the toe of the cut slope (Figure 6.8a) was not deep enough to substantially lower the water table in the vicinity of the concrete median. Although no potholes had developed but the new asphalt binder course had stripped badly and the new asphalt wearing course had started to strip at the bottom. Obviously, with time, potholes would develop due to stripping.

Most of this North-East extension section is mountainous and is predominantly built in cut areas. The subsurface drainage can be improved in this instance by

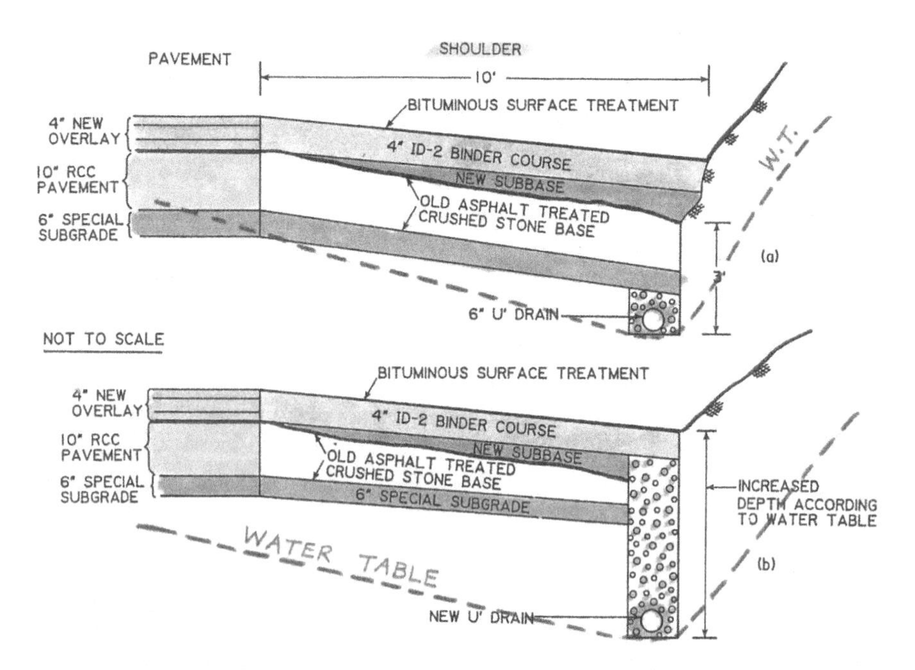

FIGURE 6.8 Typical cut section of North-East Pennsylvania Turnpike, USA.

increasing the depth of the longitudinal underdrain at the shoulder edge in cut areas. The proposed improvement as shown in Figure 6.8 will also drain the new shoulder subbase, which is sandwiched between two impermeable layers and is causing asphalt stripping in the overlying new binder course.

6.1.3 Interstate 40, Oklahoma, USA

Since the surface of a section of Interstate 40 in Oklahoma had developed extensive cracking, the asphalt pavement was milled to 90 mm depth and replaced with new asphalt binder and wearing course in October 1990. Potholes started to develop in 1991 especially in the outside wheel track of the slow (right) lane (Figures 6.9 and 6.10).

The road is at ground level (no embankment) with a high grassy median – which is real challenge for subsurface drainage. This pavement did not have any edgedrain and drainage, if any, from all pavement courses was blocked by dense shoulder. That is why potholes developed in the right wheel track of the slow (right) lane where water is likely to pond up.

The distressed road was inspected in September 1991 by excavating many pits, determining in-situ moisture content in each layer and evaluating stripping. Severe stripping was observed in the asphalt binder course which was causing potholes. Classic, textbook telltale signs of stripping in three stages can be seen in Figure 6.11. First, white spots would appear due to moisture vapor bringing up the fines from the stripped aggregate. Second, asphalt flushing would occur due to migration of asphalt binder upwards by moisture vapor. Third, potholes would develop in the flushed area due to badly stripped asphalt mix underneath.

FIGURE 6.9 Potholes on outside wheel track of Interstate 40 in Oklahoma.

6.1.4 Will Rogers Parkway, Oklahoma, USA

Work on one section of the Will Rogers Parkway was performed in June 1992. It consisted of milling 20 mm thick open graded asphalt friction course and replacing it with 25 mm thick dense graded asphalt wearing course. Potholes started to develop within one month or so in July 1992. The project was investigated in July 1992. Figure 6.11 shows the road after potholes developed. The road is almost at ground level, which is a challenge for subsurface drainage. Although the pavement has an edgedrain it was apparently not successful in draining the left half of the right, slow lane. Again, potholes developed due to stripping in the inside wheel track of the slow lane. White stains prior to development of potholes can be seen on the pavement.

No asphalt flushing was observed on the road surface after the appearance of white stains on this project. That meant that the old asphalt binder course had already lost most of the asphalt binder due to stripping. This was confirmed by excavating several pits. The old asphalt binder below the new 25 mm wearing course had severe stripping. This deteriorated, stripped asphalt course should also have been milled off. However, observing its condition was overlooked. Asphalt pavements tend to strip below open graded asphalt friction course unless suitable precautions are taken. This already badly stripped binder course was saturated with subsurface water and crumbled when sampled. That is why potholes developed within one month or so.

The preceding case histories demonstrate how subsurface water, moisture or moisture vapor can cause disintegration of asphalt mix resulting in potholing. It is very difficult to retrofit existing road pavements with subsurface drainage systems. It

FIGURE 6.10 Close-up of potholes on outside traffic lane of Interstate 40 in Oklahoma, USA

is best attained at the time the road is constructed after detailed planning and design (Federal Highway Administration, 1999).

6.2 THE IMPACT OF WATER ON THE STRUCTURAL STRENGTH OF PAVEMENTS

The impact of water on the structural strength of pavements is illustrated with a study of a failure in a National Highway (NH) in India, that occurred within a short period of time after four-laning. The subgrade, sub-base (GSB), granular base and the binder layers were constructed. However, premature failures in the form of hair line cracks and raveling were observed on the binder course. A typical pavement cross section of the failed NH-4 section is as shown in Figure 6.12. The stretch of road that had failed had observations of failures as listed in Table 6.1.

FIGURE 6.11 White stains and potholes on inside wheel track of the Will Rogers Parkway, Oklahome, USA.

The results of the tests carried out on granular sub-base layer material as part of failure investigation of the pavement are as given in Table 6.2.

It can be seen that the material used in the GSB layer contained excessive fines. The percentage of material passing through 75 micron sieve varied from 14% to as high as 17.5%. The GSB layer has to serve as a drainage layer. As the permeability of dense graded aggregate reduce with the increase in percentage of fines, it was hypothesized that the very high percentages of fines in the GSB layers have prevented efficient drainage. The high value of plasticity index has added to the problem. Both the LL and PI values are more than the specified value of 25 and 6, respectively. The percent

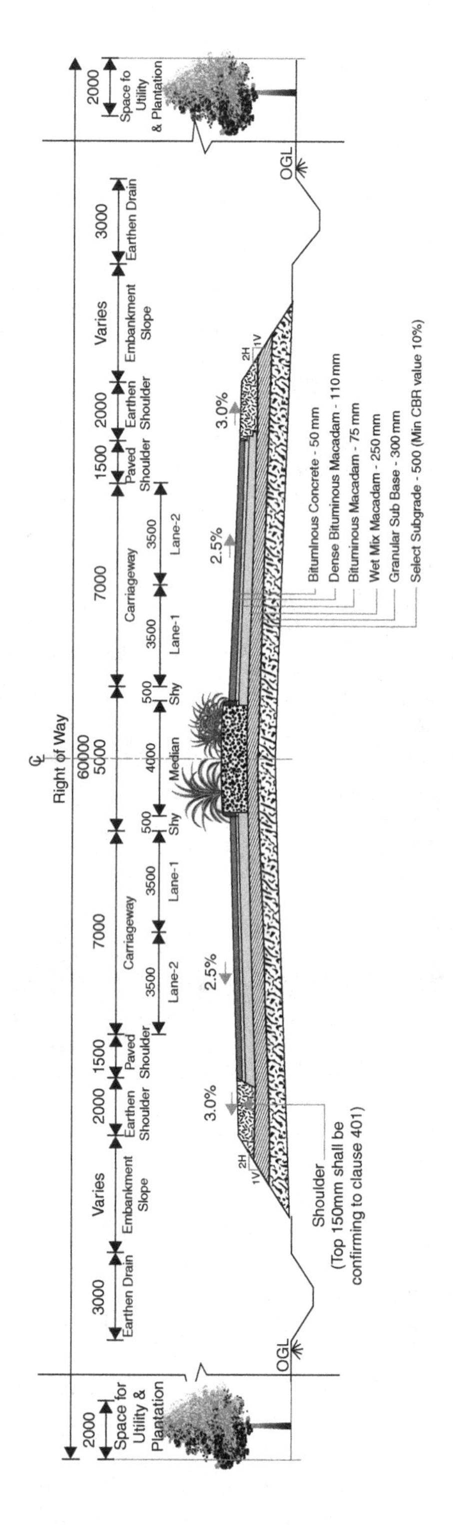

FIGURE 6.12 Typical pavement cross-section of NH-4.

TABLE 6.1
Observations of Pavement Distress Areas From Km 370–374

Station	From	To	Length (m)	Observation of Failures
1	370.470	370.510	40	Deformation and cracking of pavement near edge and median
2	370.515			Potholes
3	370.650	370.800	150	Deformation and cracking of pavement along median portion
4	370.865	371.250	385	Cracking along pavement edge
5	371.250	371.610	360	Pavement deformation
6	372.000	372.500	500	Interconnected cracks along wheel path
7	372.600	372.900	300	Interconnected cracks along wheel path
8	373.000	373.120	120	Interconnected cracks along wheel path
9	373.300	373.400	100	Interconnected cracks along wheel path
10	373.800	373.820	20	Interconnected cracks along wheel path
11	373.950	374.000	50	Interconnected cracks along wheel path

TABLE 6.2
Results of Tests Carried Out on GSB Layers

Properties	**Pavement**		**Pavement Edge**	
	Failed location	**Sound location**		
Field moisture content (%)	8.5	6.2	15.3	
Liquid limit (LL)	31	35	40	
Plasticity index (PI)	31	35	40	
Gradation of GSB				
Sieve size mm	**Specified limits**	**Fines at failed location (%)**	**Fines at sound location (%)**	**Fines at the pavement edge and shoulder (%)**
>10mm	100	100	100	100
>4.75mm	55–75	97.3	64.06	100
>0.075mm <4.75mm	10–30	56.4	36.04	92.61
<0.075mm	0–10	23.3	14.24	75.1

of fines in the drainage layer is found to vary from 14% to as high as 75% towards the pavement edge and shoulders, indicating poor drainage characteristics of the GSB layer particularly towards the shoulders, resulting in practically impervious layer. This has prevented effective drainage of water to the drains. The field moisture content of the layer used at the pavement edge was found to be as high as 15.3%. It is seen that the gradation of material used at the sound location is better than at the failed location. Hence, it can be inferred that one of the factors for the failure of this pavement is inadequate subsurface drainage in terms of poor gradation of the granular sub-base layer, higher liquid limit and plasticity index of the materials and high percentage of fines in the granular sub-base layer. Dynamic Cone Penetrometer data was used to compute the resilient modulus of the different layers, and layered elastic analyses were used to compute the maximum tensile strain at the bottom of the asphalt layer and vertical compressive strain on top of the subgrade. The values were then compared with the limiting strain values. The predicted pavement distresses were compared with the field performance to validate the obtained strain values. The monthly cumulative damage analysis was carried out according to subgrade strain criteria to study the effect of seasonal variation in moduli values on pavement performance.

The summary of calculations of resilient modulus using available DCP data for the project road was carried out and the variation in the resilient moduli values for different test pit locations is shown in Figure 6.13.

A graph showing the calculated radial tensile strains along with the laid down permissible limit, and cracking observed at various pit locations is shown in Figure 6.14.

A graph showing the calculated vertical compressive strains along with the laid down permissible limit, and rutting observed at various pit locations is shown in Figure 6.15.

The values of calculated tensile strains in asphalt layer are found to exceed the permissible tensile strain limit at all the test pit locations and cracking was observed

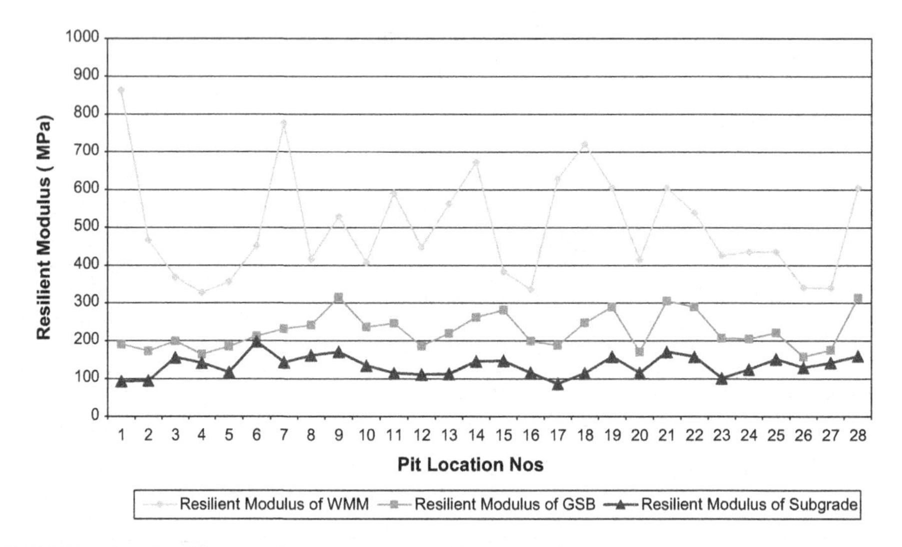

FIGURE 6.13 Resilient modulus of layers at different pit locations.

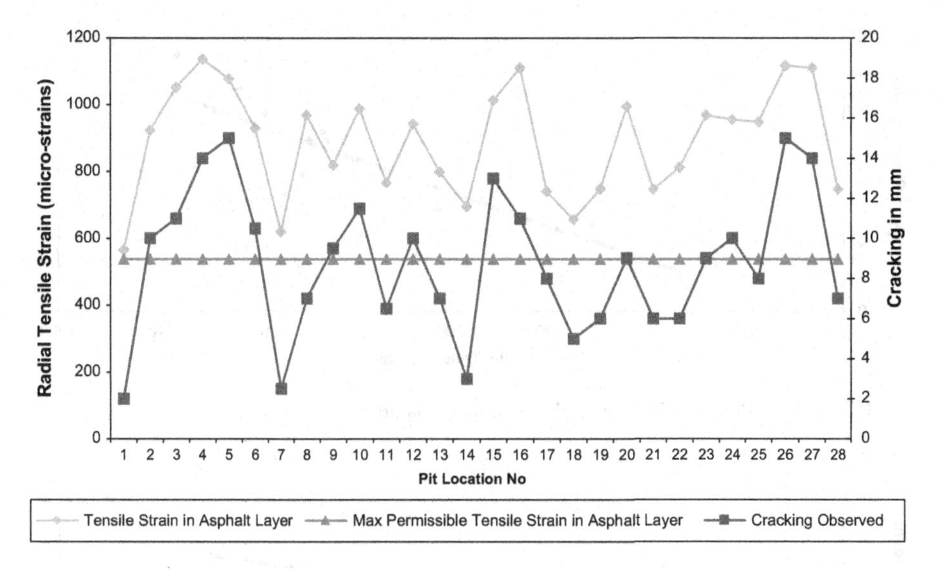

FIGURE 6.14 Calculated radial strain and observed cracking on failed stretch.

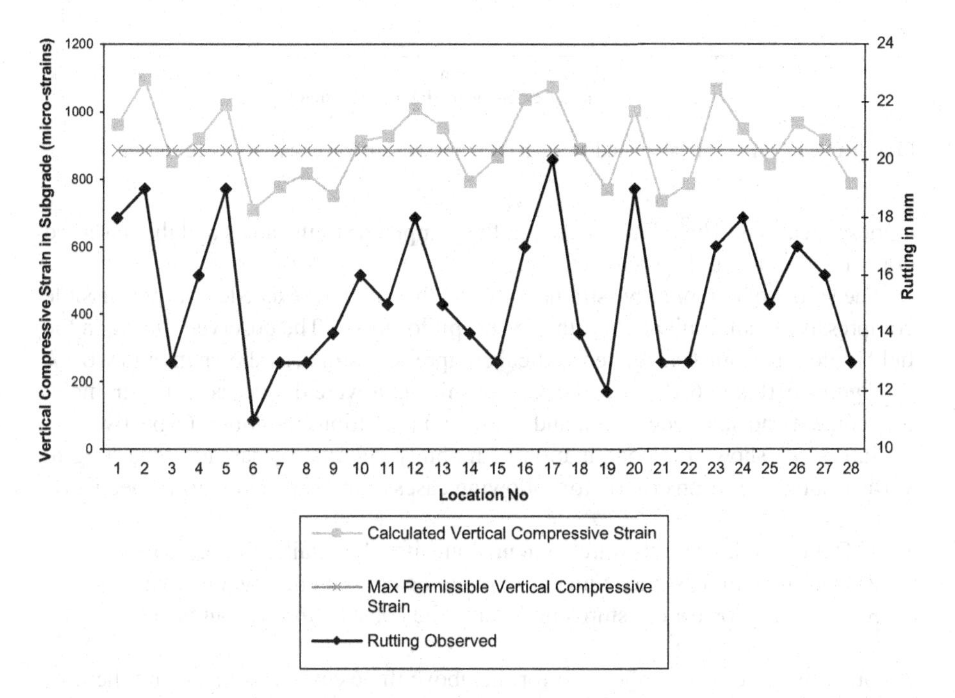

FIGURE 6.15 Calculated vertical compressive strain and observed rutting on failed stretch.

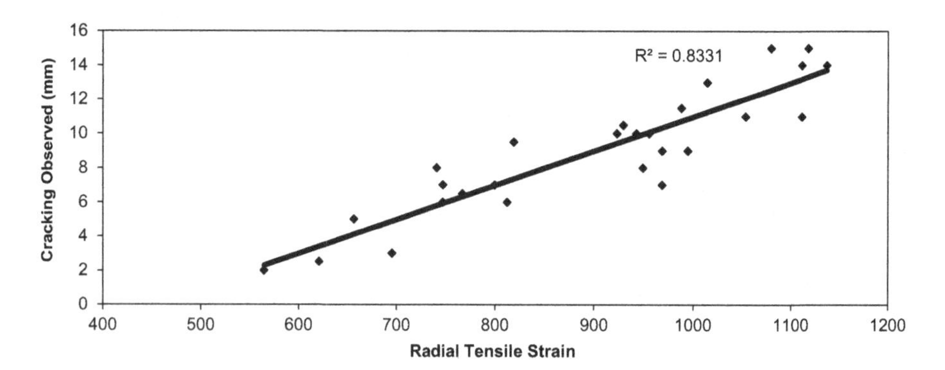

FIGURE 6.16 Plot of calculated radial strain versus observed cracking.

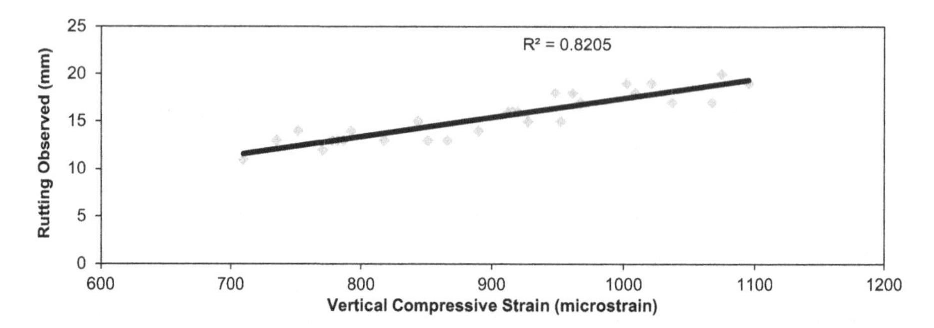

FIGURE 6.17 Plot of calculated vertical compressive strain versus observed rutting.

at these locations. The relation between the computed strain values and the observed cracking is shown in Figure 6.16.

The values of compressive strains on top of the subgrade exceeded the permissible compressive strain limit at 16 of the 28 test pit locations. The observed rutting in the field is plotted against calculated vertical compressive strain, as shown in Figure 6.17.

Figures 6.18 and 6.19 show the relationships that were developed between moisture content and subgrade strain, and strain and repetitions to failure, respectively.

Tables 6.3 shows the computation of monthly cumulative damage according to vertical subgrade strain criteria for following cases of moisture contents respectively:

1. Collected data of moisture content from site where failure observed.
2. Assuming highest moisture condition to be constant throughout year.
3. Assuming lowest moisture condition to be constant throughout year.

A comparison of the damage ratio for the above three cases was made and the analysis helps us to understand the effect of poor subsurface drainage on the performance of the pavement. Poor drainage results in a high moisture content in the pavement layers leading to a higher cumulative damage ratio.

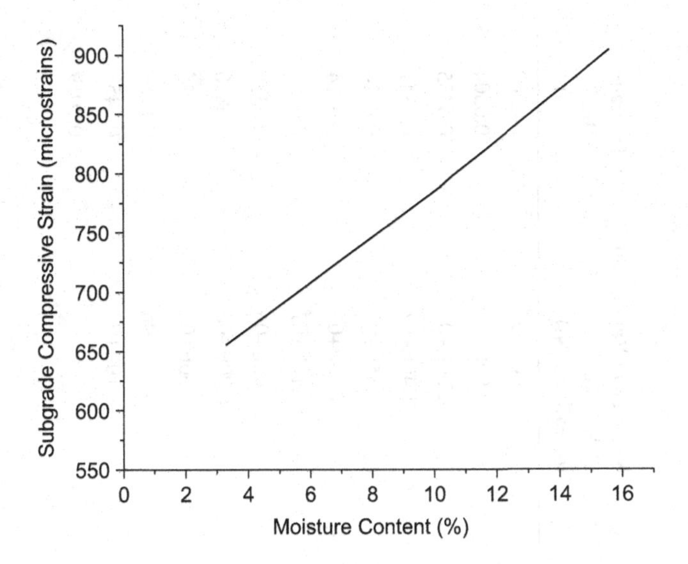

FIGURE 6.18 Plot of moisture content versus subgrade compressive strain.

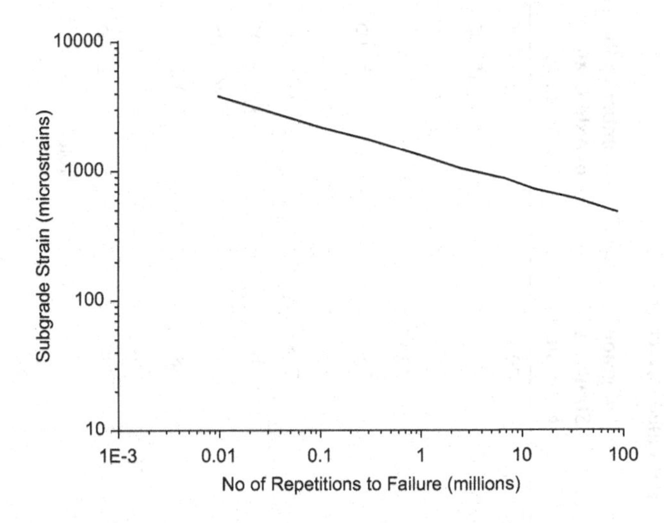

FIGURE 6.19 Sub Plot of subgrade compressive strain versus traffic repetitions to failure.

The cumulative damage ratios calculated are used to determine for all the three cases of moisture content:

1. The rut depth from relationship between rut depth and cumulative damage according to subgrade strain criterion, from Figure 6.20.
2. The cracked area index from relationship between cracked area index and cumulative damage according to subgrade strain criterion, from Figure 6.21.

TABLE 6.3
Monthly Cumulative Damage Calculations

Case 1: Collected Data from NH-4 at failed stretches

Month	Moisture Content (%)	Subgrade Strain E v Microstrains	Number of Repetition of Axle Load for the Month (N)	Number of Repetition of Axle Load for Failure (Nf)	Damage Ratio (N/Nf)
January	10.5	790	5.02E+05	1.00E+07	0.0502
February	11.1	810	4.54E+05	9.00E+06	0.0504
March	9	760	5.02E+05	1.50E+07	0.0335
April	9.1	770	4.86E+05	1.40E+07	0.0347
May	8.6	750	5.02E+05	1.55E+07	0.0324
June	8.6	750	4.86E+05	1.55E+07	0.0314
July	10.1	785	5.02E+05	1.10E+07	0.0457
August	11.6	810	5.02E+05	9.00E+06	0.0558
September	13	835	4.86E+05	6.00E+06	0.0810
October	15.3	880	5.02E+05	4.00E+06	0.1256
November	13.6	845	4.86E+05	4.20E+06	0.1157
December	10.2	780	5.02E+05	1.10E+07	0.0457
			Total Cumulative Damage =		0.7019

Case 2: Assuming Highest Moisture Condition to be Constant Throughout the Year

Month	Moisture Content (%)	Subgrade Strain E v Microstrains	Number of Repetition of Axle Load for the Month (N)	Number of Repetition of Axle Load for Failure (Nf)	Damage Ratio (N/Nf)
January	15.3	880	5.02E+05	4.00E+06	0.1256
February	15.3	880	4.54E+05	4.00E+06	0.1134
March	15.3	880	5.02E+05	4.00E+06	0.1256
April	15.3	880	4.86E+05	4.00E+06	0.1215
May	15.3	880	5.02E+05	4.00E+06	0.1256
June	15.3	880	4.86E+05	4.00E+06	0.1215
July	15.3	880	5.02E+05	4.00E+06	0.1256
August	15.3	880	5.02E+05	4.00E+06	0.1256
September	15.3	880	4.86E+05	4.00E+06	0.1215
October	15.3	880	5.02E+05	4.00E+06	0.1256
November	15.3	880	4.86E+05	4.00E+06	0.1215
December	15.3	880	5.02E+05	4.00E+06	0.1256
			Total Cumulative Damage =		**1.4783**

(continued)

TABLE 6.3
Monthly Cumulative Damage Calculations (Cont.)

Case 3: Assuming Lowest Moisture Condition to be Constant Throughout the Year

Month	Moisture Content (%)	Subgrade Strain E v microstrains	Number of Repetition of Axle Load for the Month (N)	Number of Repetition of Axle Load for Failure (Nf)	Damage Ratio (n/Nf)
January	8.6	750	5.02E+05	1.55E+07	0.0324
February	8.6	750	4.54E+05	1.55E+07	0.0293
March	8.6	750	5.02E+05	1.55E+07	0.0324
April	8.6	750	4.86E+05	1.55E+07	0.0314
May	8.6	750	5.02E+05	1.55E+07	0.0324
June	8.6	750	4.86E+05	1.55E+07	0.0314
July	8.6	750	5.02E+05	1.55E+07	0.0324
August	8.6	750	5.02E+05	1.55E+07	0.0324
September	8.6	750	4.86E+05	1.55E+07	0.0314
October	8.6	750	5.02E+05	1.55E+07	0.0324
November	8.6	750	4.86E+05	1.55E+07	0.0314
December	8.6	750	5.02E+05	1.55E+07	0.0324
			Total Cumulative Damage =		0.3815

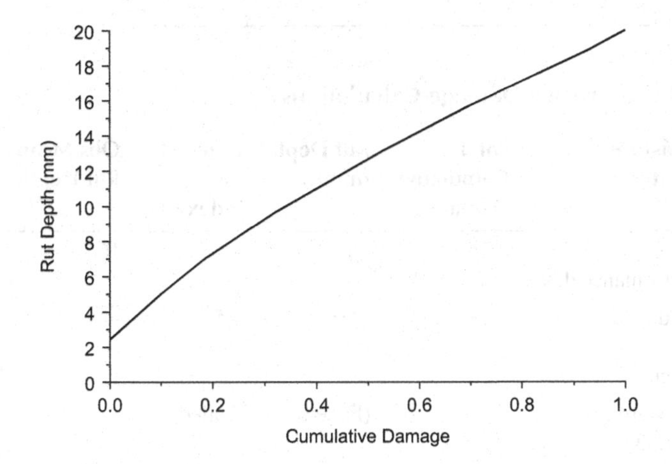

FIGURE 6.20 Plot of rut depth versus cumulative damage according to the subgrade strain criterion.

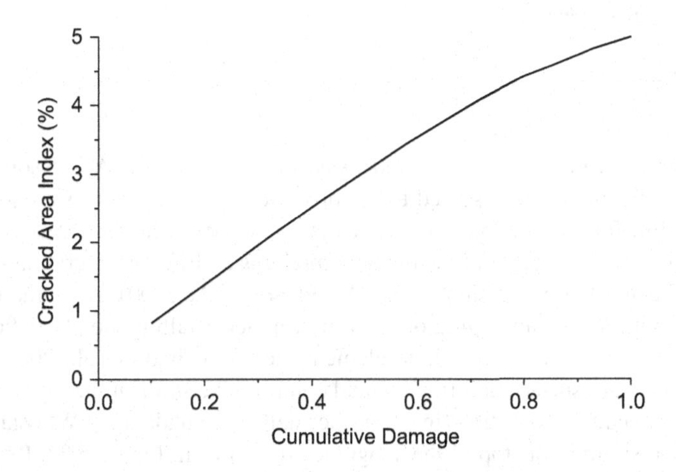

FIGURE 6.21 Plot of cracked area index versus cumulative damage according to the radial tensile strain criterion.

The summary of cumulative damage calculations and the corresponding predicted rut depth and cracked area index are shown in Table 6.4.

The above analysis showed that the failure on the stretch of National Highway was due to inadequate subsurface drainage.

6.3 FLOOD IMPACT AND ASSESSMENT OF IMPACT ON ROADS

In a study conducted by Nivedya et al., 2017, the structural condition of a three-layered pavement constructed with different base types during and after flooding was studied. The base layer varied from coarse-grained to fine-grained (by varying D60

TABLE 6.4
Results of Cumulative Damage Calculations

Case	Moisture Content	Total Cumulative Damage	Rut Depth (mm)	Cracked Area Index (%)	Obs Mean Rut Depth	Obs Mean CAI(%)
1	Field performance data	0.7	15	4	16.2	4.5
2	Predicted performance at higher	1.48	>20	> 5	–	–
	moisture content of 15.3%		(Failed)	(Failed)		
3	Predicted performance at lower	0.38	10	2.2	–	–
	moisture content of 8.6%					

from 0.1 mm to 1 mm) and the subgrade was considered to be a A-2–5 soil. The duration of the flooding was assumed to be three days. The degree of saturation of the base with time for different base types in a three-layered pavement was assessed using FE analysis. The estimation of the resilient modulus at different saturation levels was conducted using the expressions in the MEPDG (ARA, 2000). Also, the pavement responses with a varying degree of saturation under a falling weight deflectometer (FWD) was estimated using a finite element module (Tirado et al., 2007) Primary responses, such as surface deformation and compressive strain at the top of subgrade, were determined. The variations in the surface deflection under the FWD plate and the compressive strain at the top of the subgrade are shown in Figure 6.22. It was found in this study that the degree of saturation in the base layer was high and the resilient modulus dropped by around 35–40% during the flooding period.

In the study conducted by Gupta et al., 2017, various types of subgrades with different soil type classifications were considered. The effect of flooding was studied on a three-layered pavement cross section with A-6, A-4 and A-2–5 (ASTM D3282, 2015) type of subgrade. The unsaturated soil seepage analysis was carried out using FE analysis and structural strength due to saturation, was studied using layered elastic analysis (LEA). The variation in saturation and resilient modulus (M_r) of the base layer were estimated. The vertical strain on top of the subgrade, horizontal strain at the bottom of the asphalt layer and surface deflection were determined using LEA. The effects of flooding for eight days were studied and it was seen that in the case of A-6 type of soil, the modulus values did not recover even after five days of flooding whereas the A-4 type of soil shows a faster recovery; the best results were seen for A-2–5 type of soil. The damage factors were also calculated

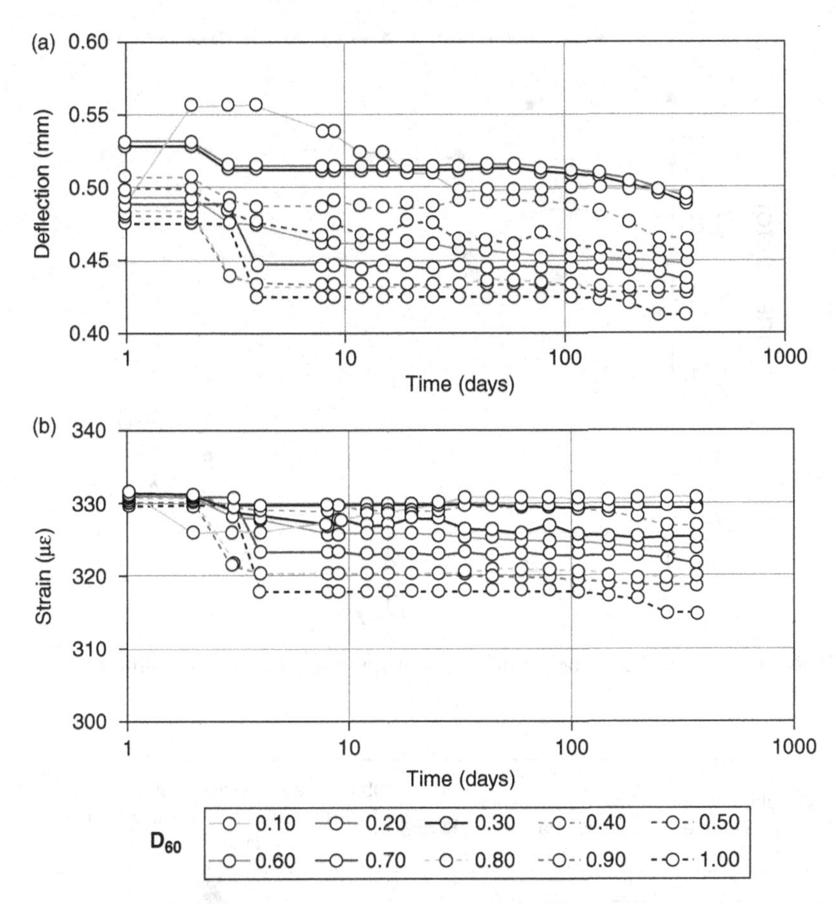

FIGURE 6.22 (a) Surface deflection; and (b) compressive strain at top of subgrade under FWD test conducted at different times after flooding for base courses with different D_{60}.

from the surface deflection values obtained from LEA. The damage factor was calculated as surface deflection in μm/500 μm (Mallick et al., 2017). When the damage factor was greater than one, it indicates that the pavement was damaged. The damage factor is found to be significantly high even after five days of flooding for the A-6 soil subgrade, whereas it drops quickly after flooding in the other types of subgrades (Figure 6.23). It was recommended that adequate drainage facilities in the form of permeable base layers combined with underdrains should be provided for pavement subgrade with A-6 type of soil.

Flood-induced moisture has caused significant structural damage to pavements. It is important to consider such damages during design and construction of pavements. In the recent years, the concept of resilience has gained popularity. Several specifications and frameworks are being developed to make the civil infrastructure components resilient to disaster induced damage. Nivedya et al. (2018) has provided a framework (Figure 6.24) for quantitatively assessing the resilience of flexible pavements to flooding.

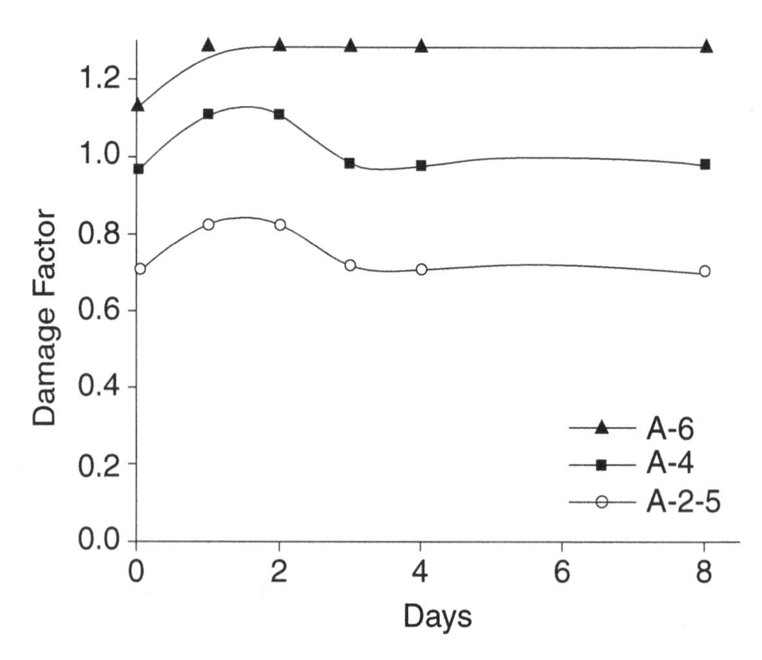

FIGURE 6.23 Damage factor for different subgrade soil types during different flooding period.

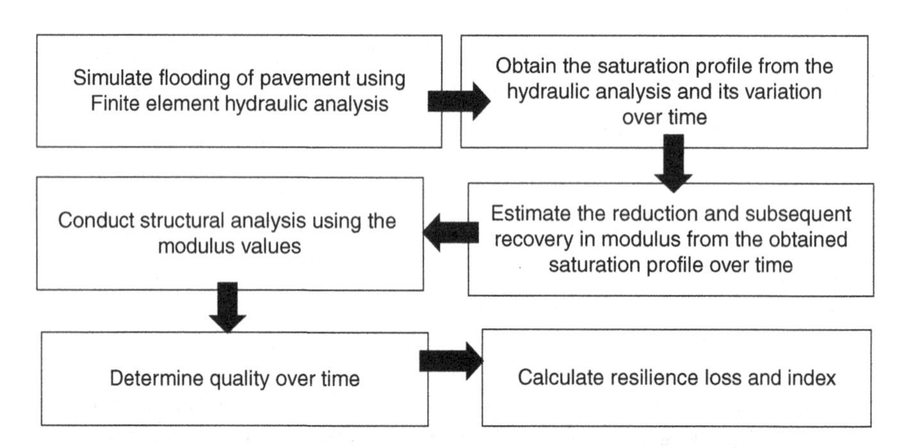

FIGURE 6.24 Framework for calculation of resilience of pavements against flooding.

The framework consists of utilizing unsaturated flow through the different layers to estimate drainage, interpretation of the results in terms of stiffness of the relevant layers, estimation of the impact of the change in stiffness on the overall structural condition of the pavement and then translating that change to a resilience index. Figure 6.25 provides an illustrative example of estimation of resilience for a pavement (Nivedya et al., 2018, Bocchini et al., 2014).

The results of a study carried out by the authors showed the need for providing base course materials with appropriate gradation to ensure adequate hydraulic

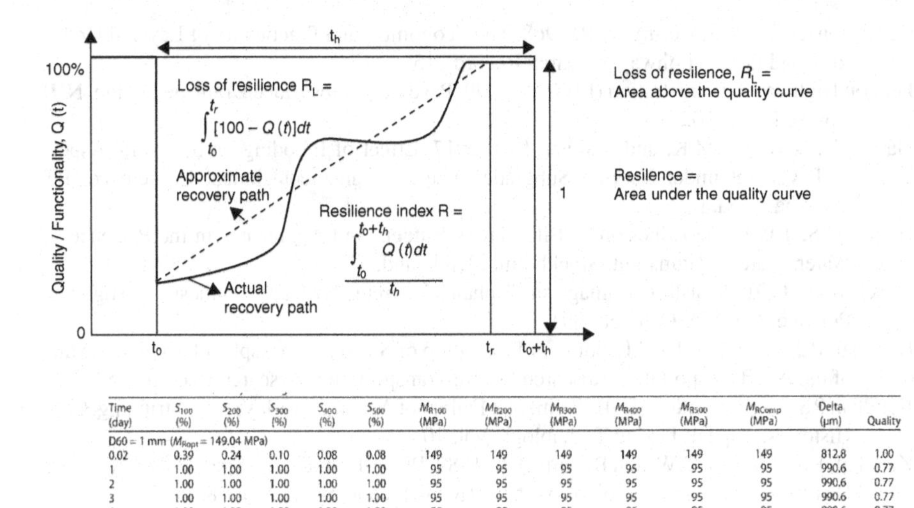

Time (day)	S_{100} (%)	S_{200} (%)	S_{300} (%)	S_{400} (%)	S_{500} (%)	M_{R100} (MPa)	M_{R200} (MPa)	M_{R300} (MPa)	M_{R400} (MPa)	M_{R500} (MPa)	M_{RComp} (MPa)	Delta (μm)	Quality
D60 = 1 mm (M_{Ropt} = 149.04 MPa)													
0.02	0.39	0.24	0.10	0.08	0.08	149	149	149	149	149	149	812.8	1.00
1	1.00	1.00	1.00	1.00	1.00	95	95	95	95	95	95	990.6	0.77
2	1.00	1.00	1.00	1.00	1.00	95	95	95	95	95	95	990.6	0.77
3	1.00	1.00	1.00	1.00	1.00	95	95	95	95	95	95	990.6	0.77
4	1.00	1.00	1.00	1.00	1.00	95	95	95	95	95	95	990.6	0.77
8.5	0.90	0.49	0.12	0.11	0.14	113	149	149	149	149	142	838.2	0.97

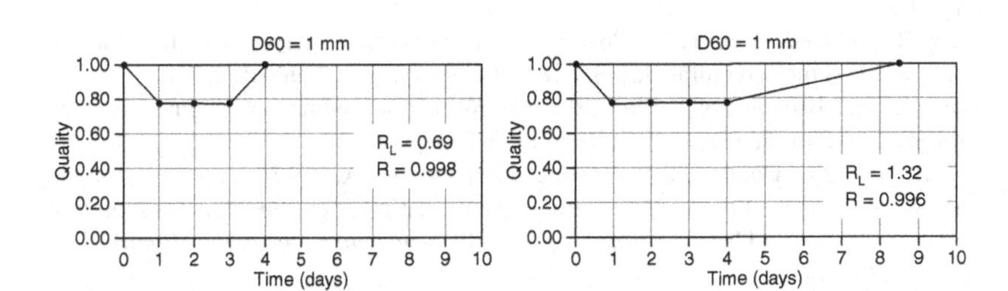

FIGURE 6.25 Illustration of estimation of resilience of pavement.

conductivity, and/or a thicker surface layer, to avoid a reduction in service quality and a loss of resilience for an extended period, in flood-prone areas.

REFERENCES

Acott, S.M. and Crawford, C. 1987. Blistering in Asphalt Pavements: Causes and Cures. National Asphalt Pavement Association, IS 97.

ARA, Inc. 2000. Appendix DD-1: Resilient Modulus as Function of Soil Moisture – Summary of Predictive Models. Guide for Mechanistic-Empirical Design of New and Rehabilitated Pavement Structures. Transportation Research Board of the National Academies, Washington, DC.

ASTM D3282-15. 2015. Standard Practice for Classification of Soils and Soil-Aggregate Mixtures for Highway Construction Purposes, ASTM International, West Conshohocken, PA.

Bocchini, P., Frangopol, D.M., Ummenhofer, T. and Zinke, T. 2014. Resilience and sustainability of civil infrastructure: toward a unified approach. *Journal of Infrastructure Systems*, 20 (2): 04014004.

Cedergren, H.R., Arman, J.A. and O'Brien, K.H. 1973. Development of Guidelines for the Design of Subsurface Drainage Systems. FHWA, Report RD-73-14, Feb. 1973.

Cedergren, H.R. and Lovering, W.R. 1968. The Economics and Practicality of Layered Drains for Road Beds. Highway Research Record 215.

Federal Highway Administration (FHWA). 1999. Pavement Subsurface Drainage Design. NHI Course No. 131026.

Gupta, A., Nivedya, M.K. and Mallick, R.B. 2017. Effect of Flooding on Structural Aspect of Pavement under Varying Subgrade Types, Indian Geotechnical Conference, IIT Guwahati, India.

Hallberg, S. 1950. The Adhesion of Bituminous Binders and Aggregates in the Presence of Water. Statens Vaginstitut, Stockholm, Meddeland, 78.

Hicks, R.G. 1991. Moisture Damage in Asphalt Concrete. NCHRP Synthesis of Highway Practice, No. 175, October 1991.

Kandhal, P.S. 1994. Field and Laboratory Evaluation of Stripping in Asphalt Pavements: State of the Art. Transportation Research Board, Transportation Research Record 1454.

Kandhal, P.S. and Rickards, I. 2001. Premature Failure of Asphalt Overlays from Stripping: Case Histories. Asphalt Paving Technology, Vol. 70.

Kandhal, P.S., Lubold, C.W. and Roberts, F.L. 1989. Water Damage to Asphalt Overlays: Case Histories. Proc. Association of Asphalt Paving Technologists, Vol. 58.

Lottman, R.P. 1971. The Moisture Mechanism that Causes Asphalt Stripping in Asphal Pavement Mixtures. University of Idaho, Moscow, Idaho, Final Report Research Project R-47, Feb. 1971.

Lovering, W.R. and Cedergren, H.R. 1962. Structural Section Drainage. Proc. International Conference on the Structural Design of Asphalt Pavements, Ann Arbor, MI.

Majidzadeh, K. and Brovold, F.N. 1966. Effect of Water on Bitumen-Aggregate Mixtures, University of Florida, Gainesville, Report CE-1, Sept. 1966.

Mallick, R.B., Tao, M., Daniel, Jacobs, J. and Veeraragavan, A. 2017. A combined model framework for asphalt pavement condition determination after flooding. Transportation Research Record (TRR): *Journal of the Transportation Research Board.* DOI 10.3141/2639-09.

Nivedya, M.K., Mallick, R.B., Tirada, C., Saremi, S.G. and Nazarian, S. 2017. Use of Artificial Neural Network to monitor pavement structural strength during adverse weather conditions, Indian Geotechnical Conference, IIT Guwahati, India.

Nivedya, M.K., Tao, M., Mallick, R.B., Daniel, J.S. and Jacobs, J.M. 2018. A framework for the assessment of resilience of pavement to flooding. *International Journal of Pavement Engineering*, https://doi.org/10.1080/10298436.2018.1533637.

Tirado, C., Carrasco, C., Nazarian, S. and Osegueda, R. 2007. Updates to Software for Estimating Damage due to Super Heavy Loads. Research Report FHWA/TX-05/9-150-01-7, Center for Transportation Infrastructure Systems, The University of Texas at El Paso.

7 Subsurface Drainage Structures – Construction and Maintenance

7.1 CONSTRUCTION OF SUBSURFACE DRAINAGE SYSTEMS

Inadequate subsurface drainage of pavement systems has contributed to many premature pavement failures. Most highway agencies agree that the presence of water is not desirable in the pavement system. However, they have different approaches to reduce the effect of water on pavements. Some agencies just completely seal the pavement including using a low permeable base with no drainage. Others would use a fully drainable pavement section with a permeable base and edgedrains, which is the primary approach discussed in detail in this chapter. There are others whose approach falls somewhere in between, for example, providing edgedrains with dense graded aggregate bases.

The primary source of water in pavements is atmospheric precipitation (rainfall). Water can enter the road pavement structural section from many sources such as:

- Cracks and/or joints in the pavement surface.
- Infiltration through the shoulder especially the joint between mainline and shoulder.
- Infiltration from the roadside ditches.
- High groundwater table or presence of aquifers.
- Capillary moisture drawn from saturated subgrade.
- Condensation of water vapor within the pavement (Figure 7.1).

A pavement subsurface drainage system is required to control and/or eliminate the ingress of water from these sources.

If there is inadequate subsurface drainage, premature pavement distresses are likely to occur. In case of concrete pavements, the following major distresses are experienced due to excess water: pumping, faulting, slab corner breaks, curling/warping,, punch outs, D-cracking, etc.

In case of asphalt (flexible) pavements, the following major distresses are experienced due to excess water: stripping (peeling away of asphalt binder film from the aggregate surface), rutting (usually due to excessive moisture in the subgrade), fatigue (alligator) cracking (due to lack of support from stripped asphalt courses or

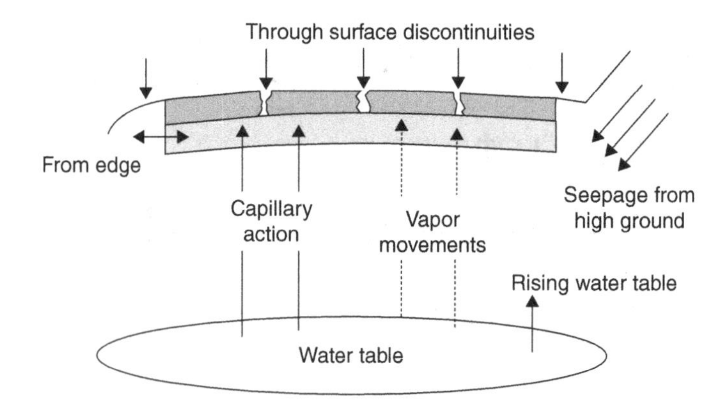

FIGURE 7.1 Sources of moisture in pavement systems. (FHWA, 2008, Federal Highway Administration. Pavement Subsurface Drainage Design. NHI Course 131026 Reference Manual. Report FHWA-NHI-08-030. Figure 2.1.)

saturated subgrade), potholes (due to localized disintegration of asphalt courses from stripping), etc.

Considerably high ingress of rainwater can occur through the unsealed pavement cracks/joints especially the main pavement/shoulder joint as compared to that from the subgrade. Therefore, it is highly desirable to keep the cracks/joints properly sealed at all times. Full width paving to eliminate the lane-shoulder joint is quite helpful. However, some water would still infiltrate. Thus, to minimize or eliminate the premature distresses mentioned above, it is necessary to have a good pavement subsurface drainage system.

In the past, subbases placed directly below the concrete pavements, or placed within the asphalt (flexible) pavements, were usually dense graded and not free-draining. This did not present any significant problems because the traffic was low both in terms of volume and intensity. Now, with increased traffic, all the problems mentioned above have come to the fore and need to be dealt with.

Lovering and Cedergren (1962) recognized, during the 1960s, that a rapid and "positive" subsurface drainage system was needed to take care of water infiltration into the pavement from the subgrade and from the pavement surface. They proposed the use of an open-graded aggregate drainage layer directly below the concrete or asphalt pavement constructed on subgrades or subbases containing few fines. It was recommended to treat the open-graded aggregate with asphalt binder in case a more stable permeable base was required. This was termed asphalt treated permeable base (ATPB). Portland cement can also be used in lieu of asphalt binder. Figures 7.2 and 7.3 show such a system for asphalt pavements and concrete pavements, respectively (FHWA, 2008).

For subgrades/subbases containing a larger percentage of fines, a filter type aggregate drainage layer was recommended below the ATPB. The so-called two-layer drainage system was considered to provide rapid, positive drainage to water infiltration from the surface as well as from the subgrade. If the subbase has reasonable

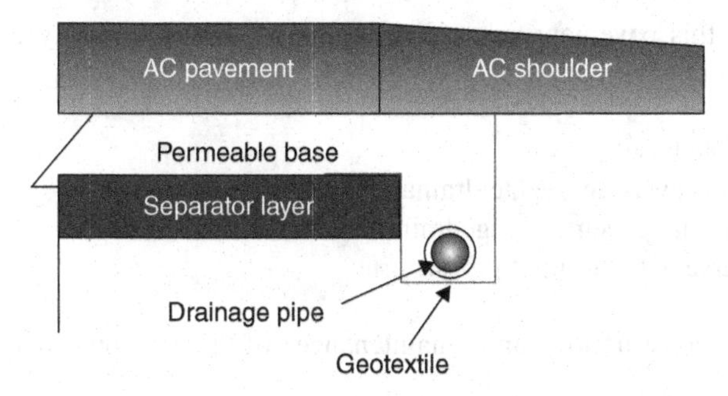

FIGURE 7.2 Asphalt pavement structure with a permeable base. (FHWA, 2008, Federal Highway Administration. Pavement Subsurface Drainage Design. NHI Course 131026 Reference Manual. Report FHWA-NHI-08-030.)

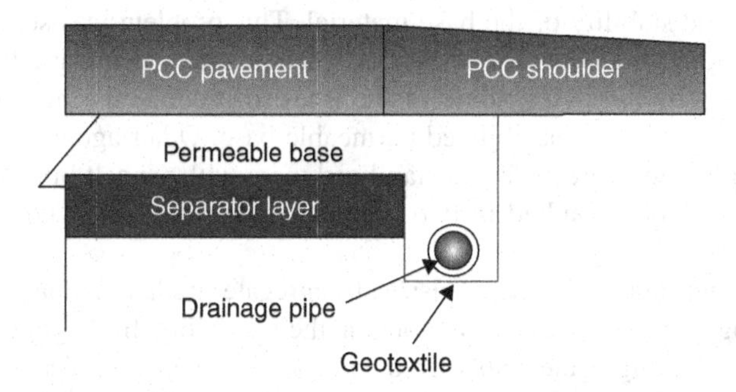

FIGURE 7.3 Concrete pavement structure with a permeable base. (FHWA, 2008, Federal Highway Administration. Pavement Subsurface Drainage Design. NHI Course 131026 Reference Manual. Report FHWA-NHI-08-030.)

permeability, geotextile can be used in lieu of the filter type aggregate drainage layer. The geotextile would prevent migration of fines from the subbase to ATPB (Lovering and Cedergren, 1962).

For the drainage layer to be effective, a sufficient drainage outlet with an adequate flow capacity has to be provided at the pavement edge. This consists of a longitudinal, collector pavement edgedrain and associated transverse outlet pipes at reasonable spacing.

Cedergren and Lovering (1968) had emphasized that this whole longitudinal and transverse drainage outlet system should be maintained well otherwise its clogging would create an undesirable "bath tub" effect in the permeable base. It is not recommended to daylight the permeable base layer because it is likely to be clogged from roadway debris, vegetation and lawn mowing. In the case of shallow ditches, storm water may also flood the drainage layer.

This pavement subsurface drainage system was first implemented by the California Department of Transportation (CALTRANS) during the mid-1980s followed by other states in the US later.

Therefore, this pavement subsurface drainage system primarily consists of the following four components:

1. Permeable base.
2. Separator layer (aggregate drainage layer or geotextile).
3. Longitudinal pavement edgedrain.
4. Transverse, lateral outlet pipes.

The design, construction and maintenance of these four components are discussed below.

7.1.1 Permeable Bases

It has been recognized that permeable base designs must be a careful balance of permeability and stability of the base material. This problem was solved with two approaches. Some highway agencies used their existing dense-graded aggregate base gradations by removing some of the fines to produce the necessary permeability and used it as an unstabilized permeable base. Other agencies used readily available single size aggregates and stabilized them with asphalt binder or Portland cement. These efforts resulted in two types of permeable bases: unstabilized and stabilized.

Therefore, unstabilized bases consist of aggregate gradations that contain some finer sized aggregates. These bases obtain their stability from good mechanical interlocking of the aggregates. Stabilized bases are more open-graded and thus much more permeable. The cementing action of the stabilizer material, at the point of aggregate contact develops the necessary stability.

The combination of base thickness and permeability must be capable of handling the design flows and keeping the saturation time to a minimum. In most cases, a thickness of 100 mm has been found to be adequate for permeable bases. Draining 50% of the drainable water in one hour is recommended as a criterion for the highest-class highways, while draining 50% of the drainable water within two hours is recommended for most Interstate highways and roads (FHWA, 1992). Detailed hydraulic calculations for determining the water flow into the permeable base are given in the US Federal Highway Administration (FHWA) Reference Manual on Pavement Subsurface Drainage Design. A Windows-based computer program *Drainage Requirements in Pavements*, version 2.0 (DRIP2) is also available to conduct the hydraulic design of permeable bases, separator layer, and longitudinal edgedrains (FHWA, 2008).

7.1.2 Unstabilized Permeable Bases

An unstabilized permeable base consisting of non-stabilized aggregates is placed over a previously constructed aggregate or geotextile separator layer. Aggregate material should be hard, durable and angular. The preference is 100% crushed aggregate. Los Angeles abrasion should not exceed 45%. The soundness percentage loss should not exceed 12% or 18% as determined by the sodium sulfate test and magnesium sulfate

TABLE 7.1
Gradation Requirement for Unstabilized Permeable Base

Sieve Size (mm)	Percent Passing
37.5	100
25	95–100
12.5	60–80
4.75	40–55
2.36	5–25
1.18	0–8
0.30	0–5

test, respectively (FHWA, 2002). The aggregate should conform to the gradation requirement given in Table 7.1.

The gradation in Table 7.1 can generally be obtained by a 50:50 blend of AASHTO Gradations Nos. 57 and 9.

The aggregate is placed with an asphalt laydown machine in compacted lift not exceeding 100 mm. Compaction should consist of one to three passes of a 4.5 to 9 metric ton smooth steel-wheeled power roller. Over rolling can cause degradation of the aggregate material resulting in loss of permeability. Permeability should be at least 2,000 feet (610 m) per day.

7.1.3 Asphalt Stabilized Permeable Bases

It is most commonly called asphalt treated permeable base (ATPB) in the US. The quality requirements for aggregate should be the same as those for an unstabilized permeable base.

The grade of asphalt binder (performance grade, viscosity grade or penetration grade) should be the same as that used for overlying asphalt courses. The asphalt binder content should be 3 +/– 0.5% by mass of dry aggregate. Although, AASHTO Gradation 57 or 67 can be used for gradation, the gradation used by New Jersey Department of Transportation as given in Table 7.2 has been recommended by the Federal Highway Administration (FHWA, 2002). AASHTO Aggregate Gradations 57 or 67 (Table 7.3) have also been used by some highway agencies.

An antistripping agent (liquid or hydrated lime) must be used in the asphalt mix. The ATPB material should be spread at a temperature between 93 and 120°C (Figure 7.4). Compaction should begin when the mix has cooled to 65°C and should be completed before the temperature falls below 38°C. Compaction should consist of one to three passes of a 4.5 to 9 metric ton smooth steel-wheeled power roller. Over rolling can cause degradation of the aggregate material resulting in loss of permeability. Permeability should be at least 2000 feet (610 m) per day.

Compacted thickness of ATPB is usually 100 mm. Most highway agencies do not assign any structural strength to the ATPB while conducting pavement structural design.

TABLE 7.2
Gradation Requirements for Asphalt Treated Permeable Base (ATPB)

Sieve Size (mm)	Percent Passing
25.0	100
19.0	95–100
12.5	85–100
4.75	15–25
2.36	2–10
0.075	2–5

TABLE 7.3
Gradation Requirement for Cement Stabilized Permeable Base

Sieve Size (mm)	AASHTO 57 Percent Passing	AASHTO 67 Percent Passing
37.5	100	–
25.0	95–100	100
19.0	–	90–100
12.5	25–60	–
9.5	–	20–55
4.75	0–10	15–25
2.36	0–5	2–10
0.075	–	2–5

7.1.4 Cement Stabilized Permeable Bases

The quality requirements for aggregate should be the same as those for an unstabilized permeable base. The gradation of aggregate should meet either AASHTO No. 57 aggregate or AASHTO No. 67 aggregate gradation. Both gradations are given in Table 7.3.

Cement should be Type I, Type I-p or Type II conforming to ASTM C 1950–97 Portland cement. An application rate of 112–167 kg/cu m (2–3 bags per cu yd) is recommended. Curing compound should be white pigmented wax base concrete curing compound meeting the requirements of ASTM C 309.

Standard concrete paving techniques should be followed for placing and consolidating the cement treated aggregate permeable base material. Permeability of the completed permeable base should be at least 2,000 feet (610 m) per day.

FIGURE 7.4 Placing asphalt treated permeable base (ATPB). (FHWA, 2008, Federal Highway Administration. Pavement Subsurface Drainage Design. NHI Course 131026 Reference Manual. Report FHWA-NHI-08-030.)

7.2 SEPARATOR LAYER

A separator layer must be provided between the permeable base and the subbase/subgrade to keep subgrade soil particles from contaminating the permeable base (Figures 7.2 and 7.3). A separator layer over stabilized subbases/subgrades may not be needed provided the stabilized material is not subjected to saturation or high pressures for an extended period of time. A separator layer can be provided by an aggregate separator layer or geotextile. The Federal Highway Administration (FHWA) recommends the use of an aggregate separator layer due to its ability to spread vertical loads over a wide area of the subgrade.

7.2.1 Aggregate Separator Layer

The aggregate separator layer must perform many important functions:

1. The aggregate separator layer must be stable to provide a firm working platform for constructing the permeable base without rutting or movement during the paving operation. Since most highway agencies use a dense graded aggregate base for the aggregate separator layer, it is generally strong enough to support the paving operation.

TABLE 7.4
Gradation of Aggregate Separator Layer

Sieve Size (mm)	Percent Passing
37.5	100
19.0	55–90
4.75	25–60
0.300	5–25
0.075	3–12

2. The gradation of the aggregate separator layer must be carefully selected so that fines are not pumped up from the subgrade into the permeable base. Basic aggregate filtration equations are used to establish the particle size and gradation of the aggregate separator to prevent contamination of the permeable base. Such calculations are given in the US Federal Highway Administration (FHWA) Reference Manual on Pavement Subsurface Drainage Design (FHWA, 2008).
3. The aggregate separator layer should have a low permeability so that it can act as a shield to deflect infiltrated water over to the edgedrain (FHWA, 2002).

A typical gradation, as used by the New Jersey Department of Transportation, for dense graded aggregate separator layer is given in Table 7.4.

The aggregate separator layer should consist of durable, crushed, angular aggregate material with good mechanical interlocking characteristics. The aggregate for the separator layer should at least meet the requirements for a Class C Aggregate in accordance with AASHTO M 283 for Coarse Aggregate for Highway and Airport Construction. The material should be non-plastic.

The maximum density of the compacted aggregate separator material should be determined in a laboratory by AASHTO T 99, Moisture Density Relationship Using A5-lb (2.5 kg) Hammer and an 12 in (305 mm) Drop, Method D, or AASHTO T 180–97, Moisture Density Relationship Using a 10 lb (4.54 kg) Hammer and an 18 inch (457 mm) Drop, Method D.

Compaction of the aggregate separator layer should continue until the material complies with the compaction acceptance criteria: in situ density should be at least 95% of the maximum dry density. Water should be applied uniformly over the materials during compaction in the amount necessary to obtain the required density.

7.2.2 Geotextile Separator Layer

Geotextiles have been used to replace graded granular filters in drainage applications. They must allow water to flow through the filter into the permeable base; they must retain soil particles in place and prevent their migration into the permeable base; and they should not blind or clog.

Geotextiles used for subsurface drainage should meet AASHTO M 288 standard. The requirements include: grab strength; sewn seam strength; tear strength; puncture strength; burst strength, permittivity; apparent opening size; and ultraviolet stability. AASHTO specification gives permittivity (sec^{-1}) and apparent opening size (mm) of the geotextile based on the percent in situ material passing 0.075 mm sieve in the layer below the permeable base (FHWA, 2002).

Successive sheets of geotextile should be overlapped a minimum of 300 mm. Atmospheric exposure of geotextiles to the elements following laydown should be a maximum of 14 days to minimize damage potential.

7.3 LONGITUDINAL PAVEMENT EDGEDRAINS

Longitudinal edgedrains play a key role in drainable pavement systems. The edgedrain must have the necessary hydraulic capacity to handle the water being discharged from the permeable base. As the water moves towards the outlet each element of the drainage system should increase in capacity. The runoff that enters the pavement section through the permeable base should drain quickly to the edgedrain. Depending on the cross slope of the road, longitudinal edgedrains may be needed on both sides of the road pavement. If the pavement slopes to one side only, that side should receive the edgedrain.

Longitudinal pavement edgedrains have been retrofitted to existing concrete and asphalt pavements which did not have permeable bases. Performance results in such cases have been mixed. However, in most cases the retrofitting at least takes care of the water infiltrating through the pavement/shoulder joint.

According to the US Federal Highway Administration (FHWA), geocomposite fin drains are not recommended with permeable bases due to their low hydraulic capacity; durability concerns; and placement problems.

7.4 TRENCH AND PIPE

The trench backfill and edgedrain pipe must have the necessary capacity to handle the design flows. The trench backfill material should have the same permeability as the permeable base course to ensure capacity. The geotextile used to wrap the edgedrain trench should not extend up into the permeable base to block the flow from it.

Conventional pipe edgedrains are recommended because of their relatively high flow capacity and their ability to be maintained easily. Pipes must be strong enough to resist the loads placed on them. There are many different plastic materials used in the manufacture of plastic pipes. Most highway agencies use flexible, corrugated polyethylene (CPE) or smooth, rigid, polyvinyl chloride (PVC) pipe. Pipe should conform to appropriate AASHTO specification.

The pipe should have 42.32 sq cm of openings per linear meter (2 sq in.ft) of pipe to allow the discharge water to flow into the edgedrain. Trench backfill material should be stable and at least as permeable as the permeable base material.

Depending on pipe size, many highway agencies use a trench width of 200–250 mm (8–10 inches). The trench width must be wide enough to facilitate proper placement of the pipe and compaction of the backfill material around the pipe.

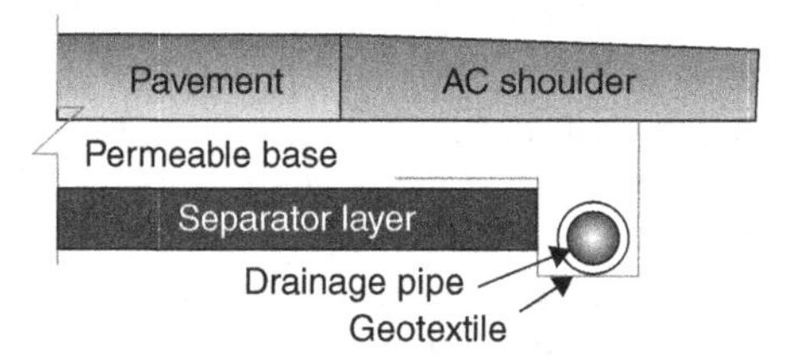

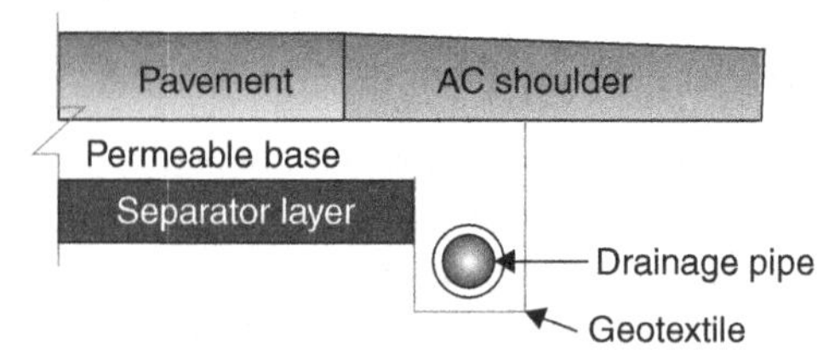

FIGURE 7.5 Location of edgedrain: pre-pave and post-pave installation. (FHWA, 2008, Federal Highway Administration. Pavement Subsurface Drainage Design. NHI Course 131026 Reference Manual. Report FHWA-NHI-08-030.)

The trench depth must be adequate to accomplish the intended drainage function. It is recommended that the trench depth be deep enough to allow the top of the pipe to be located 50 mm (2 inches) below the bottom of the permeable base as mentioned earlier (FHWA, 2002).

The trench backfill should be compacted adequately to prevent premature settlement of the shoulder. Trench backfill should be placed in at least three layers with compaction being applied after each lift.

The edgedrain trench should be lined with a geotextile to prevent the fines from the surrounding subgrade from entering the pavement section. Thus, the primary purpose of the geotextile is filtration; thereby keeping the fines in the subgrade from contaminating the trench backfill material. The geotextile should have permeability several times greater than the subgrade soils.

As mentioned earlier, detailed hydraulic calculations for the flow in edgedrain; trench backfill; and geotextile around the trench are given in the US Federal Highway Administration (FHWA) Reference Manual on Pavement Subsurface Drainage Design (FHWA, 2008).

The location of longitudinal edgedrains in terms of lateral and vertical placement is discussed below for concrete and asphalt pavements. Figure 7.5 shows a typical location of an edgedrain both for pre-pave and post-pave installations.

7.4.1 Location for Concrete Pavements

As far as the vertical placement of the edgedrain is concerned, the top of the edgedrain pipe should be located 50 mm (2 inches) below the bottom of the permeable base so that water can drain out of the permeable base into the edgedrain pipe.

For concrete pavements with asphalt shoulders, the pavement–shoulder joint usually opens up allowing a relatively large amount of water to enter the pavement section at this point. The edgedrain should be located as close to this joint as feasible. This will provide a short, direct path for the water to pass through the permeable base to the edgedrain pipe. Since the edgedrain is located next to the edge of the concrete pavement, time to drain will be kept to a minimum. Also, since the edgedrain is located under the paved shoulder, the paving will provide protection for the pipe from heavy wheel loads.

7.4.2 Location for Asphalt Pavements

The lateral and vertical placement of edgedrains for asphalt pavements is quite similar to that stated for concrete pavements. Most likely any asphalt pavement would have an asphalt concrete shoulder and asphalt stabilized permeable base.

Again, for vertical placement of the edgedrain, the top of the edgedrain pipe should be located at least 50 mm (2 inches) below the bottom of the permeable base so that water can drain out of the permeable base into the edgedrain pipe. Since drainage of the entire flexible pavement section is warranted, the edgedrain should be located as low as possible. Obviously, this will be controlled by the invert of the roadside ditch.

For lateral placement of the edgedrain, the edgedrain should be located as close to the main pavement as possible. This will provide a minimum time to drain for the permeable base under the main pavement. By locating the edgedrain approximately 30 inches (0.7 meters) from the main pavement, some drainage of the shoulder will occur. Since the edgedrain is located under the paved shoulder, the paving will provide protection for the pipe from heavy traffic loads.

7.5 TRANSVERSE, LATERAL OUTLET PIPES

Installation of the outlet pipe is critical to the drainage system because it drains to the longitudinal edgedrain. Use of a high stiffness polyethylene or PVC non-perforated pipe is recommended for the outlet pipe to ensure proper grade and sufficient stiffness to withstand installation and traffic (mowers) loads without damage or significant deformation. The outlet pipe should have a minimum stiffness of 65 psi. Plastic pipe conforming to ASTM D 3034–89 Type PSM Poly (Vinyl Chloride) (PVC) Sewer Pipe and Fittings with a Standard Diameter Ratio (SDR) of 23.5 or ASTM D 2665, Poly (Vinyl Chloride) (PVC) Plastic Drain and Vent Pipe and Fittings, Schedule 40 are the most common rigid plastic pipes produced at the present time (FHWA, 2002).

It is recommended to provide a 3% positive downward slope of the pipe to the roadside ditch. This will ensure that the pipe will drain even if there is a slight variance of the pipe grade.

Subsurface drainage design must be coordinated with surface drainage. The invert of the outlet pipe should be at least 150 mm (6 inches) above the ten-year design flow in the roadside ditch as shown in Figure 7.6.

In the past, the outlet pipes have been connected to the mainline edgedrain using 100 mm by 100 mm (4 inch by 4 inch) tees. This design is not desirable because maintenance and video inspection equipment cannot traverse this turn. The edgedrain pipe

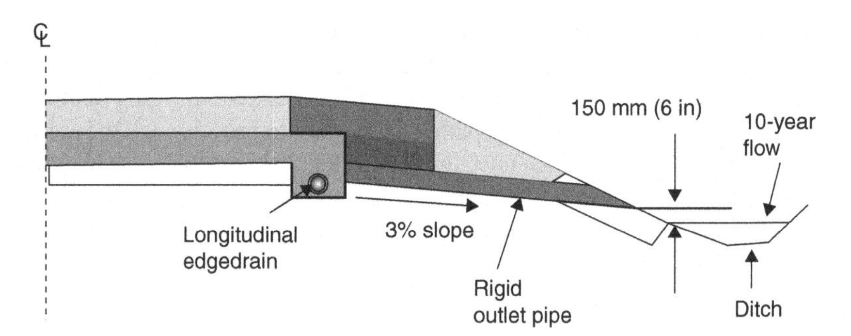

FIGURE 7.6 Recommended design for edgedrain outlet. (FHWA, 2008, Federal Highway Administration. Pavement Subsurface Drainage Design. NHI Course 131026 Reference Manual. Report FHWA-NHI-08-030.)

system should be designed with maintenance in mind. An edgedrain system with outlet pipes at both ends of an edgedrain system should be provided.

Smooth, long radius bends (typically with a radius of 0.6–1.0 m) should be provided in the edgedrain system so that jet rodding or cleaning equipment can be used easily.

Since the purpose of subsurface drainage is to remove water as soon as possible, outlet pipe spacing should be limited to 76 m (250 ft) based mainly on maintenance considerations. The edgedrain should be segmented so each segment drains independently.

7.6 HEADWALLS

Headwalls are recommended because they provide the following functions: (a) protect outlet pipe from damage by mowing operations; (b) prevent slope erosion by spreading out the water flow; and (c) locate outlet pipe for maintenance purposes (FHWA, 2002).

The most important function of a headwall is to locate the pipe outlet. If the outlet pipe cannot be found, there will be no maintenance of the edgedrain system. Therefore, large headwalls are the best type of reference markers.

Headwalls should be placed flush with the slope so that mowing operations are not impaired. Headwalls can consist of both cast-in-place and precast concrete.

Rodents have been reported to damage geocomposite fin drains and they build nests in pipe edgedrains. Therefore, rodent screens should be installed. Eroded fines can build up on the screen and plug the outlet. Rodent screens should be easily removable so that the screens and outlet pipes can be cleaned. Figure 7.7 shows a typical headwall design for dual outlet systems.

7.7 VIDEO INSPECTION

It is recommended to video record the completed edgedrain with closed circuit video equipment for the final acceptance of the project. This action would also improve the

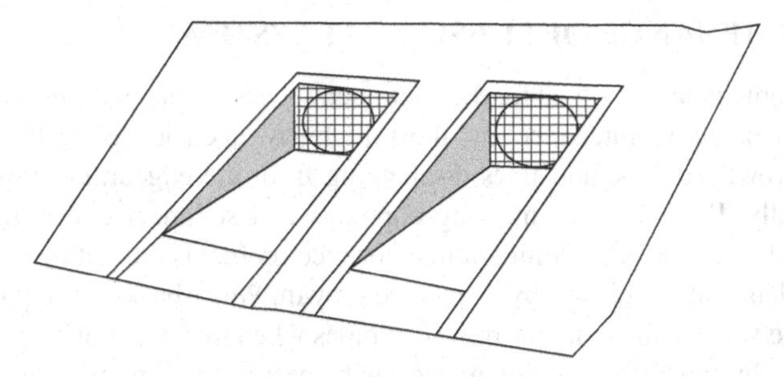

FIGURE 7.7 Recommended headwall design for a dual outlet system. (FHWA, 2008, Federal Highway Administration. Pavement Subsurface Drainage Design. NHI Course 131026 Reference Manual. Report FHWA-NHI-08-030.)

FIGURE 7.8 Inspection video camera for pipe edgedrains. (Daleiden, 1998; FHWA, 2008, Federal Highway Administration. Pavement Subsurface Drainage Design. NHI Course 131026 Reference Manual. Report FHWA-NHI-08-030.)

quality of both design and construction. Excavations and inspections of edgedrains have shown that many edgedrain pipe sections have been crushed or clogged.

The video equipment consists of the following three main units: camera head, camera control unit, and a push rod which is coiled on a metal reel. The camera used is Pearpoint flexible high resolution, high sensitivity, waterproof, color video camera. Six high intensity lights provide the necessary illumination inside the pipes. The picture is transmitted back to the control module using a closed-circuit TV system. A distance counter within the system can accurately locate the problem, if any (Figure 7.8).

7.8 MAINTENANCE OF SUBSURFACE SYSTEM

Although maintenance is critical for the continued success of any longitudinal edgedrain system, inadequate maintenance by most highway agencies is really a problem. Vegetative growth, debris, and fines discharging from the edgedrains, plug the outlet pipe eventually. Rodent nests, mowing clippings and sediment collecting on rodent screens at headwalls are also common maintenance problems. Sometimes, it is difficult to locate outlets that are hidden by vegetative growth. Water backs up in plugged outlet pipes, and the water will gush out from such pipes when they were unplugged.

It is quite obvious that no maintenance can be performed if maintenance personnel cannot find the outlets. If concrete headwalls, reference markers or painted arrows on the shoulders are used, providing maintenance becomes easy. Maintenance activities should not block or damage the pipe outlet.

Flushing or jet rodding the system is important in the maintenance scheme. Therefore, proper bends would facilitate this operation. The edgedrain pipe system should be designed with maintenance in mind. These operations should be done on a routine schedule.

Edgedrain outlets and pipe systems should be inspected at least once a year to determine their condition. As mentioned earlier, use of video equipment to inspect the edgedrain pipe system is recommended. Flushing of the pipe systems should be performed as necessary. Mowing around the outlet pipes should be done at least twice a year to keep vegetative build up to a minimum.

Based on field experience, most of the problems are found in the lateral outlet pipes which need to be cleared. Problems in the main longitudinal pipes are not very common (FHWA, 2002).

The following important conclusions have been drawn from studies on the maintenance of highway edgedrains (Christopher and McGuffey, 1997):

- The cost of maintenance is far outweighed by the anticipated design life of the road that comes with edgedrains that perform.
- There is a significant cost in terms of poor performing pavements to agencies that use edgedrains and do not have an effective preventive maintenance program.
- Long term maintenance is essential to obtain the anticipated benefits of drainable pavement systems.

If a highway agency is not willing to make a maintenance commitment, permeable bases should not be used since the pavement section may get flooded and would increase the rate of pavement damage (FHWA, 1992).

7.9 PAST EXPERIENCE WITH PERFORMANCE OF ASPHALT TREATED PERMEABLE BASES (ATPB)

Although CALTRANS pioneered the use of asphalt treated permeable bases (ATPB) during the mid-1960s, the highway agency experienced the following problems on earlier projects (Harvey et al., 1999):

- Stripping of asphalt binder from the aggregate especially in areas below the cracks/joints of the overlying pavement. This was considered to be a contributor to faulting of concrete pavements and loss of structural strength of asphalt pavements.
- Presence of unsealed cracks/joints in the surface significantly increased the infiltration rate of water into the ATPB.
- Infiltration of water into the ATPB was also occurring through overlying asphalt layers which were not compacted adequately.
- Intrusion of fines into the ATPB from the underlying aggregate subbase was occurring.
- Clogging of longitudinal and outlet pipes was observed on some projects.

It was observed that CALTRANS did not follow all the recommendations made by Lovering and Cedergren (1962) in earlier ATPB projects. For example, lower asphalt content, typically 1.5% by mass of aggregate, was used and no antistripping agent was used. This probably caused premature stripping of asphalt mixes. Also, only 25% of the aggregate particles used in the ATPB were crushed, which did not provide adequate stability.

The guidelines and recommendations for permeable bases and associated drainage structures given earlier in this chapter have addressed the issues experienced by CALTRANS which has also made the necessary changes. If these guidelines are followed, permeable bases should provide rapid and positive subsurface drainage for both concrete and asphalt pavements. There is significant evidence that a combination of good sealing (to keep the surface water from infiltrating the pavement); good positive drainage; and commitment to long-term maintenance of the subsurface drainage system, will lead to optimum performance of the pavement system.

7.10 FILTER DESIGN

Empirical relationships such as those relating the D15 of the filter material, and D15 and D85 of the base/soil material are often used for the design of filters. The criteria are as follows:

$$D_{15f} \leq 4D_{85b} \text{ (from soil-retention criterion)} \quad (7.1)$$

$$D15_{f} \geq 4D_{15b} \text{ (from permeability criterion)} \quad (7.2)$$

Filter criteria design guidelines are also provided by the US Department of Agriculture (Babu and Srivastava, 2014) as shown in Table 7.5.

Using the condition of soil boiling as the basis, Srivastava and Babu (2011) has developed an analytical approach by considering soil properties, representative particle sizes from both soil/base and the filter and hydraulic conditions for recommending filter selection criteria. They developed the factor of safety from both soil retention and permeability criteria. For the soil retention condition, an analytical expression for the critical hydraulic gradient was developed, and for the permeability criterion a factor safety expression was developed from the consideration of

TABLE 7.5
Filter Design Criteria

Base Soil Category	% Finer Than No. 200 (0.075 mm) after Regrading where Applicable	Base Soil Description	Filtering Criteria (max D_{15f})	Permeability Criterion (min D_{15f})
1	> 85	Fine silts and clays	≤ 9 but not less than 0.2 mm	For all categories, $\geq 4 \times D_{15b}$ before regrading, but not less than 0.1 mm
2	40–85	Sands, silts, clays, and silty and clayey sands	≤ 0.7 mm	
3	15–39	Silty and clayey sands and gravel	$\leq \left(\frac{40-A}{40-15}\right)[(4 \times D_{85b} - 0.7\,mm)] + 0.7\,mm$ Where A = % passing No. 200 sieve after regrading (if 4 × D85b is less than 0.7 mm, use 0.7 mm)	
4	<15	Sands and gravel	$\leq 4 \times D_{85b}$ after regrading	

Source: Babu and Srivastava, 2014.

representative particle size, thickness of the filter and the base layers and density and specific gravity of soils. The expressions are as follows.

Critical hydraulic gradient from soil retention criterion:

$$(i_{cr})_m = \frac{\left[\frac{d_s^3}{6} + \frac{1}{2\sqrt{2}} d_s^3 \tan\phi\right](G_s - 1)}{\left[\frac{3}{32} d_s n_f D_{of}^2 + \frac{1}{4} d_s^3\right]} \tag{7.3}$$

$$D_{ef} = \frac{1}{\sum \frac{\Delta S_i}{D_i}} \tag{7.4}$$

$$D_{of} = 2.67 \frac{n_f}{1 - n_f} \cdot \frac{D_{ef}}{\alpha} \tag{7.5}$$

Factor of safety from soil retention criterion:

$$(FS)_m = \frac{\left[\frac{2d_s^3}{3} + d_s^3 \tan\phi\right](G_s - 1)}{\left[\frac{3}{8} d_s n_f D_{of}^2 + d_s^3\right] i} \tag{7.6}$$

ϕ = Angle of internal friction of base soil
d_s = Weighted average particle size of the base soil
n_f = Porosity of filter media
D_{of} = Equivalent diameter of the pore channels in the filter medium

Factor of safety from permeability criterion:

$$FS_p = \frac{(G_s - 1)}{m\,p\,H\,i}\left[\left(\frac{D_{15f}}{D_{15b}}\right)^2 \frac{(mp-1)^3}{m(p-1)^3} H_b + H_f\right] \tag{7.7}$$

Where, the parameters are defined as follows:

$$p = \frac{G_s \gamma_w}{(\gamma_d)_b} \tag{7.8}$$

$$m = \frac{(\gamma_d)_b}{(\gamma_d)_f} \tag{7.9}$$

$(\gamma_d)_b$ = dry unit weight of base soil
$(\gamma_d)_f$ = dry unit weight of filter medium

7.11 GEOCOMPOSITE DRAIN

For the proper and quick drainage of the subsurface water the permeable base course is a necessity. Note that this layer is bi-functional – it serves as an effective drainage channel as well as a structural layer (just underneath the surface layer). To achieve the desirable permeability this base course needs to be constructed with a coarse and a gap gradation, which has been found to compromise its structural strength. Although asphalt or cements stabilized base courses can be utilized (as discussed earlier), in those cases the permeability gets significantly compromised.

One relatively recent solution to this problem is the use of geocomposites, which consists of two layers of planer geotextiles (acting as filters) with a drainage core in between, as shown in Figure 7.9.

It is a triplanar structure in which the external filters consist of needles punched or thermally bonded non-woven geotextiles and a drainage core or geonet that consists of a structure of sets of parallel ribs overlaid and integrally connected with a rhomboidal shape. The geonet is made up of high-density polyethylene, which is stabilized by carbon black.

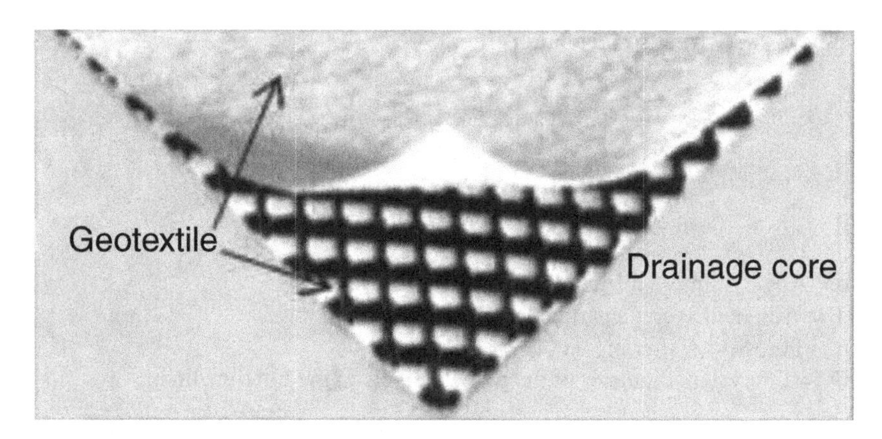

FIGURE 7.9 Geocomposite structure.

The advantages of geocomposites are in the reduction of thickness of the pavement and the avoidance of a low structural strength base course. Geocomposites provide excellent drainage as well as structural strength, and the geotextiles in the geocomposites drain are more efficient as filters compared to aggregate filters. The use of geocomposites can reduce the life cycle cost of the project.

The equivalent permeability of the geocomposite drain can be calculated as follows.

$$k_{eq} = \frac{1}{H_{gsb} + H_{gt}} [\tau_{gtr} + k_{gsb} H_{gsb}] \tag{7.10}$$

Where,

k_{eq} is the equivalent permeability;
H_{gsb} is the height of granular subbase
H_{gt} is the thickness of geotextile
k_{gsb} is the saturated coefficient of permeability of granular subbase
τ_{gtr} is the residual transmissivity of the geotextile (at 250 kPa normal stress)

Note that the transmissivity decreases with an increase in the normal stress (see Figure 7.10 as an example).

An example of reduction in thickness of the base layer (and hence the total thickness of the pavement) for different transmissivity values are shown in Figure 7.11. Note that the reduction becomes more significant when the permeability of the aggregate base layer is low which can be the case if the base course gets clogged with finer particles due to the relatively less efficiency of aggregate filters compared to geotextiles.

Li et al. (2017) conducted a field and laboratory study to compare the drainage characteristics of PCC, Warm Mix Asphalt (WMA) and unpaved section using geocomposite drainage layers placed directly underneath the surface layer. They utilized a large-scale laboratory horizontal permeameter to estimate the horizontal permeability, and core hole and air permeameter test (using the PPT) to determine the in-place permeability. FWD tests were also conducted to estimate the combined elastic modulus to evaluate the change in stiffness from the incorporation of the

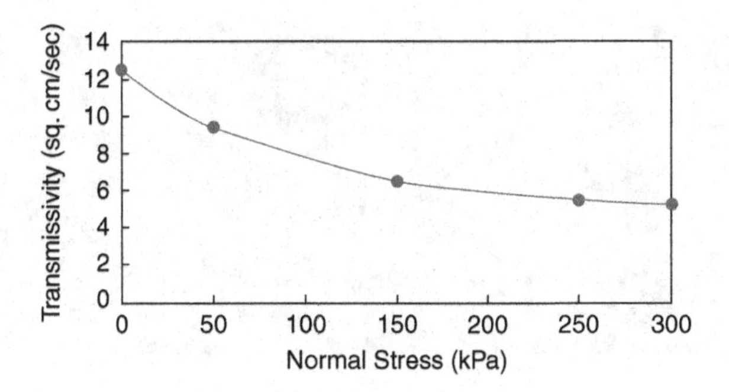

FIGURE 7.10 Plot of transmissivity versus normal stress.

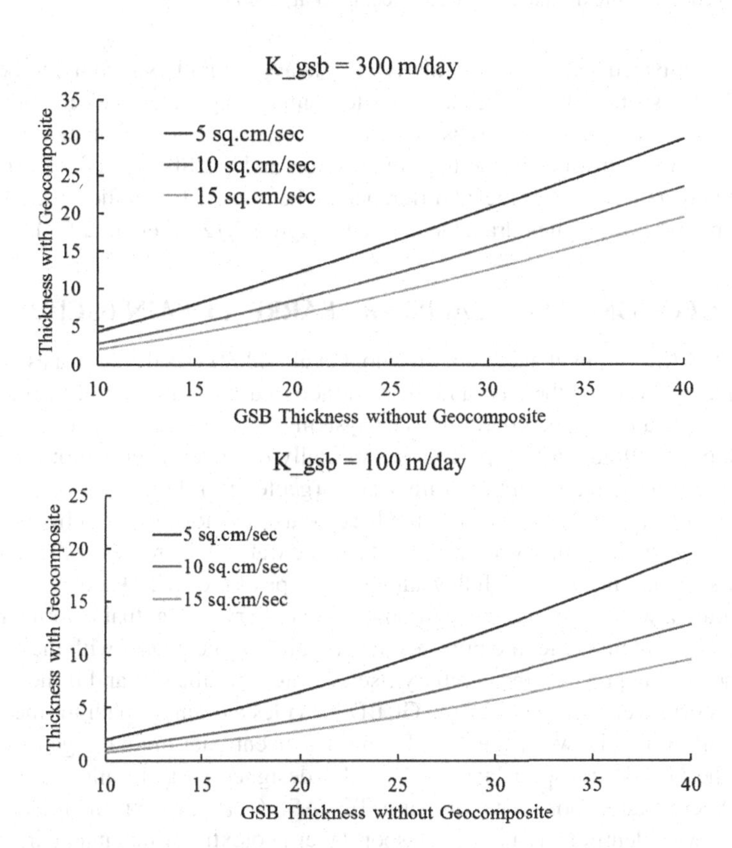

FIGURE 7.11 Savings in thickness of base course with geocomposite.

geocomposite drainage layers. The geocomposite drainage layers consisted of a synthetic polymer geonet core between two layers of non-woven geotextiles. The authors concluded that the geocomposite drainage layers improved and uniformed drainage conditions significantly for all of the pavement sections, and also resulted in a higher stiffness for the unpaved section. They also noted that the use of geocomposite (with a cost of construction that is higher than that of pavements without geocomposite) can be cost effective since it provided cost savings after 11 years of service. The ride

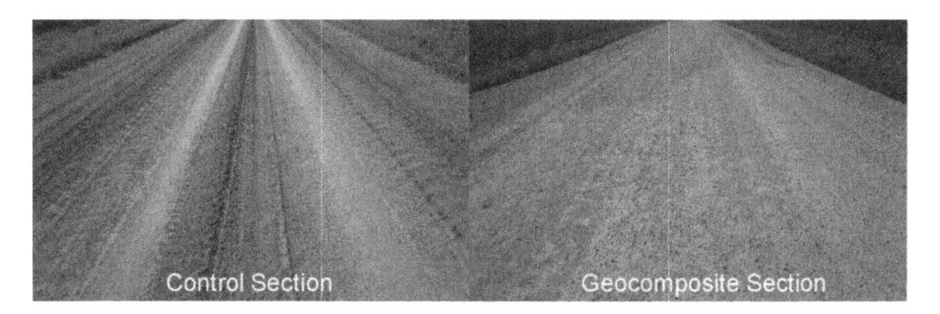

FIGURE 7.12 Comparison of conditions of unpaved section after rainfall with and and without geocomposite drainage layer. (Source: Li et al, 2017)

quality of unpaved sections, particularly after heavy rainfall, was found to be much better for the section with the geocomposite drainage layer than the section without it. They concluded that the provision of a 4% slope was not adequate for the full drainage of the rainwater in the unpaved section and that the use of geocomposite drainage layers made a significant difference in the ride quality of the pavement after rainfall by providing quick drainage of water (Figure 7.12, Li et al., 2017).

7.12 GEOCOMPOSITE CAPILLARY BARRIER DRAIN (GCBD)

The GCBD (Stormont et al., 2006; Stormont et al., 2009) was developed as a system that is placed between the base and the subgrade to drain water out of the pavement system under unsaturated condition. The system consists of three layers as follows (from top to bottom): a transport layer (specially designed geotextile); a capillary barrier (geonet); and a separator or filter layer (geotextile) (Figure 7.13).

The geonet prevents the water in the base course to move to the subgrade, while the upper geotextile (transport layer) conducts the water (k increases as the saturation increases). Water in a slope will flow along the slope of the GCBD and as long as it is not saturated, water does not move through the capillary barrier that is formed by the geonet. The bottom geotextile prevents the clogging of the geonet with the subgrade soil. The GCBD prevents the capillary rise of water into the soil and if the base and the transport layer gets saturated the GCBD provides a drainage path for the water.

The difference between the GCBD and a conventional drainage geocomposite is that the GCBD transport layer is specially designed so as to allow drainage of water under negative pore water pressure. TGLASS (heavy, woven and multifilament material) was identified as a good transport layer geotextile which maintained transmissivity over a large range of suction.

A study conducted at the MNRoad test facility (see Figure 7.14) proved the efficiency of GCBD during storm events. The study concluded that the function of the GCBD is dependent on the climate and has drainage benefits under low saturation in all climates. It indicate the following benefits: reduced equilibrium water content in the base course; prevention of positive pore pressure in the base course; reduction of increase in moisture content of subgrade due to infiltration of water from the surface; and prevention of capillary rise of water from the subgrade to the base.

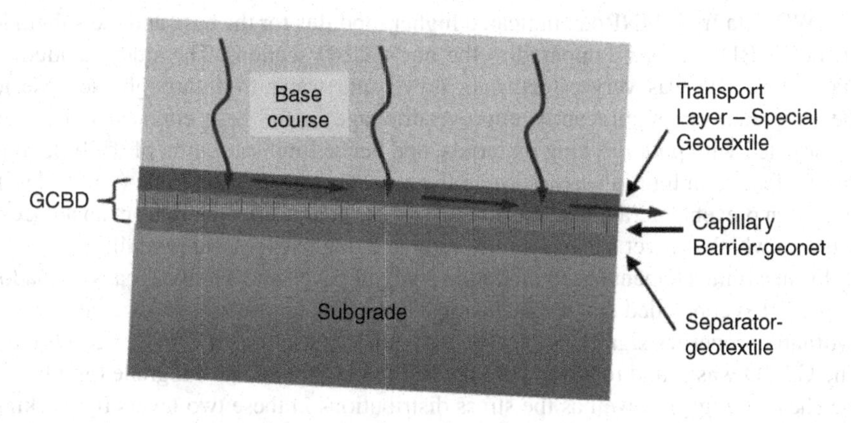

FIGURE 7.13 Concept of GCBD. (Source: Stormont, J.C., Henry, K. and Robertson, R. 2006. Geocomposite Capillary Barrier Drain (GCBD) for Limiting Moisture Changes in Pavements: Product application. www.researchgate.net/publication/241868376. Used with permission.)

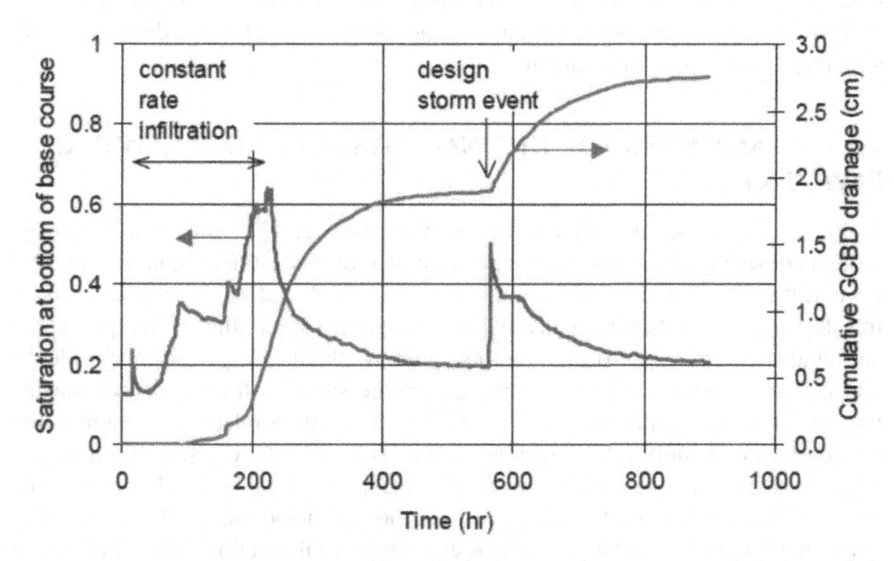

FIGURE 7.14 Plot of field data indicating the effectiveness of GCBD in preventing saturation of the base course. (Source: Stormont, J.C., Henry, K. and Robertson, R. 2006. Geocomposite Capillary Barrier Drain (GCBD) for Limiting Moisture Changes in Pavements: Product application. www.researchgate.net/publication/241868376. Used with Permission.)

The cost of the GCBD was reported to be $20/m^2 at the time of the writing of this report (2009). The installation of a GCBD involved tying it to the edgedrain collection system.

FWD data from MNRoad indicated higher modulus for the base and the subgrade of the GCBD sections compared to the non-GCBD sections. The study conducted that the GCBD was very effective in preventing water flow through the base to the subgrade, can significantly improve drainage in the base courses if they are constructed with poor draining materials, and hence limit saturation of the base, it is more effective in limiting saturation in the case of shallow water table (1 m or less), and it can prevent lateral flow and isolate the base course effectively from an adjacent soil area which can get highly or fully saturated due to excessive rainfall.

From a finite element based modeling study of paved and unpaved roads, Bahador et al. (2013) concluded that the inclusion of a GCBD improves both mechanical and hydraulic properties significantly, especially under unsaturated moisture conditions. The GCBD was found to affect pore pressure in both base and subgrade (and hence the shear strength) as well as the stress distributions in these two layers by working as a reinforcement layer. They noted that the GCBD increased the suction in the (silty sand) subgrade by up to 8 kPa and decreased the suction in the aggregate base course by up to 3.6 kPa during simulated rainfall (constant flux of 19mm/h, to simulate 6-h rainfall with a 50-year return period for Greensboro, NC, USA). They mentioned that the GCBD with the woven fiberglass geotextile as the transport layer caused 2 kPa less reduction in suction in the base course compared to the nonwoven polypropylene geotextile because of the higher Water Entry Value (WEV) of the WF geotextile. The hydraulic benefits were increased and the mechanical benefits were decreased with an increase in the pavement thickness.

7.13 FRAMEWORK FOR DRAINAGE SYSTEM UNDER NEW PCC PAVEMENT

Mallela et al. (2000) provided a framework for designing a subsurface drainage system consisting of a permeable base for a new or reconstructed jointed concrete pavement (JCP). Some of the important factors that should be considered to make this decision are site factors, which include: subgrade type; traffic level; topography; and climatic factors; and design factors, such as: shoulder type; dowels; widened lanes; joints sealants; and PCC durability. Some states in the US use permeable base under all new concrete pavements (California), others relate it to design traffic (Wisconsin) or a multitude of factors such as subgrade soil class, traffic load and volume, pavement type and functional classification (Minnesota). The factors that are significant with respect to the ranking of the site include subgrade permeability (for bottom drainage), site topography (cut, at-grade or fill section) and climate (wet-nonfreeze, wet-freeze, dry-nonfreeze and dry-freeze). The design ranking is affected by traffic (design traffic and functional class of the pavement), type of shoulder (tied PCC or other) and other design factors such as presence of dowel at transverse joints and widened lanes.

Rankings from the above two considerations are combined to produce a recommendation for or against a subsurface drainage system (see Table 7.6).

Note that the permeable base should be considered as an option once the drainage system is recommended. For "feasible" cases local experience should be taken into

TABLE 7.6
Considerations for The Determination of Need for Permeable Base System

a) Ranking of Site Conditions Based on Climate and Subgrade

Subgrade Condition		Climatic Condition			
		Non-Freeze		Freeze	
		Dry	Wet	Dry	Wet
Permeability >30 m/day	Fill section	Good	Good	Fair	Poor
	Cut section	Good	Fair	Fair	Poor
Permeability 3–30 m/day	Fill section	Good	Fair	Poor	Poor
	Cut section	Fair	Fair	Poor	Poor
Permeability <3 m/day	Fill section	Fair	Poor	Poor	Poor
	Cut section	Fair	Poor	Poor	Poor

b) Ranking of Traffic and Design Conditions

Type of Shoulder		Traffic Condition		
		Low (<4.5 Million ESALs; 20-year design lane)	Medium (4.5–20 million ESALs)	Heavy (>20 million ESALs)
Tied PCC/ widened lane	Dowels	Good	Good	Fair
	No dowels	Fair	Poor	Poor
Other	Dowels	Good	Fair	Poor
	No dowels	Poor	Poor	Poor

c) Determination of Need for Subsurface Drainage

		Site Rankings		
		Good	Fair	Poor
Traffic and design rankings	Good	Not recommended		Feasible*
	Fair	Not recommended	Feasible*	Recommended
	Poor	Feasible*	Recommended	Recommended

Note: Consider: 1) past pavement performance; 2) past experience; 3) anticipated quality of aggregate used for paving; 4) in-place water table and any modification of gradeline to improve drainage; 5) anticipated increase in service life while using various drainage alternatives; 6) cost differential while using various drainage alternatives.

Source: Mallela et al., 2000.

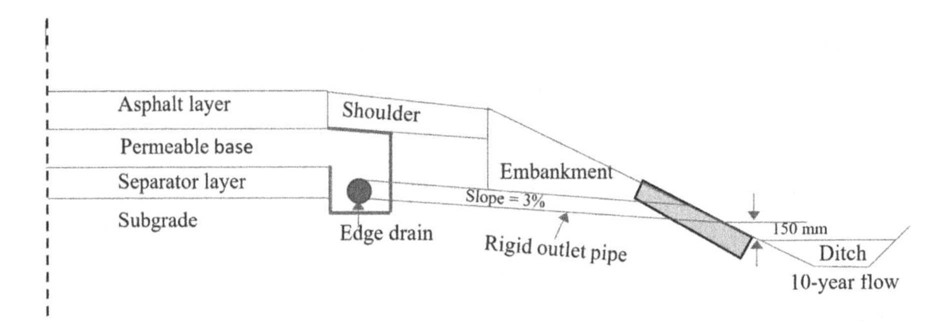

FIGURE 7.15 Permeable base system. (redrawn from Mallela et al., 2000).

account to decide, and the final decision should be based on a consideration of life-cycle cost. For pavements with concrete that is susceptible to "D" cracking failure, a permeable base is definitely required.

A schematic of the permeable base drainage system is shown in Figure 7.15, where the minimum laboratory permeability of the permeable base material is 300 m/day. A rigid outlet pipe system should be attached to the permeable base and longitudinal edgedrain system to carry the outflow to a ditch or water body that can accommodate a ten-year flow.

The authors recommended all crushed aggregate for the permeable base, and either cement stabilized (Portland cement treated base, PCTB) or asphalt stabilized (permeable asphalt-treated, PATB) material for high type pavements. They recommended a minimum of 3% AC 40-type asphalt binder with a modified Lottman's test value of >85% for PATB. The PCTB should have a cement application rate of 130–170 kg/m^2.

Stabilized or unstabilized dense graded aggregate layer can be used for the separator or the filter layer whose function is to prevent the ingress of the subgrade material into the permeable base material or vice versa, provide an impermeable barrier to route the water from the permeable base horizontally towards the edge and to serve as a construction platform for base and surface layers. For critical design conditions (consider site as well as traffic), the use of geotextile fabric between the base and the separator/filter layer is recommended.

The edgedrain consists of longitudinal pipes placed at 50 mm from the bottom of the trench on the side of the pavement at the lane-shoulder joint. The function is to collect the water coming out of the pavement and transmit it to the outlet. Both pipe and prefabricated geocomposite edgedrain are recommended for edgedrain – the pipe can be of smooth walled polyvinyl chloride (PVC). The edgedrain trench should be backfilled with stabilized or unstabilized open graded materials with a minimum permeability equal to the permeability of the base. The trench needs to be stable enough to prevent the settlement of shoulder and deterioration. The trench should be lined with geotextile in all areas except the part that faces the permeable base.

The outlet pipe carries the water from the edgedrain to the ditches or water bodies. These are short pipes made of metal non-perforated and smooth and rigid with rodent screens and headwalls with a 3% slope, and designed to minimize clogging over the long term.

The side ditches (or storm drain in urban areas) receive water from the collector pipes. They should have adequate freeboard and a minimum longitudinal grade of 0.005 m/m.

An alternative design is the daylighting of the permeable base without any edgedrain or outlet, especially for areas with very low longitudinal grade.

The hydraulic design of the permeable base system consists of the design of the geometry, permeable base material and thickness, edgedrain and trench and outlet pipe and ditch components. For the geometry, pavements are expected to have minimum longitudinal and transverse cross slopes to allow positive drainage. If the minimum longitudinal slope cannot be maintained, then the special transverse drain should be installed.

The hydraulic design of permeable base can be done with consideration of either depth of flow or time to drain. The latter is preferred because the depth of flow requires the calculation of infiltration amount into the pavement and have been found to result in conservative estimates of thickness of permeable base course. The time to drain method is based on the combination of specific period of time to drain a specific percentage of water (degree of saturation) considering a 100% saturated condition as a starting point. For JPCP the recommendation is to provide 100 mm of permeable base with materials with a minimum permeability of 300 mm/day. AASHTO classification of drainage quality is based on the time to drain from 100% to 50% level of saturation: excellent = <2 hours; good = <1 day; fair = <7 days; and poor = >30 days. High type pavements should be designed with excellent drainage systems.

The edgedrain pipe should have size and spacing to maintain hydraulic capacities that exceed the expected discharge from the permeable base, and the pipe should be greater than or equal to 100 mm in diameter for allowing a video camera inspection and maintenance. Consideration for maintenance should be made to fix the edgedrain pipe diameter and outlet pipe spacing and spacing of the outlet pipe should not exceed 75 m to allow for effective cleaning.

For stability of the permeable base, a higher C_u (coefficient of uniformity) is preferred, provided the material is well graded. $C_u = D_{60}/D_{10}$; D_{60} is the sieve size that corresponds to 60% passing the gradation plot, D_{10} is the sieve size corresponding to 10% passing in the gradation plot. C_z, coefficient of gradation (ASTM D2487, $D30^2/(D10*D60)$), can be used to "qualify" the material as well graded or not. For well graded materials, $C_u > 4$, $C_z = 1–3$. Closer the C_z is to 1, more uniform is the material (See Figure 7.16). The Corps of Engineers (COE) specification is $C_u > 3.5$, $C_z = 0.9–4.0$.

Mallela et al. (2000) recommended the use of Cu >4 and if the criteria cannot be met (such as AASHTO #57 or #67) or if additional strength is required then the layer should be stabilized with suitable material. They recommended a value of Cz = 0.6–1.6 (see Table 7.7 for list of permeable base properties).

A sensitivity plot (Figure 7.17, redrawn from data from Mallela et al., 2000) of various factors for the design PCC slab thickness illustrates that the drainage coefficient (Cd, AASHTO 93) is one of the most significant factors – change in 10% causes a change in 12.5 mm in PCC slab thickness. Note that any Cd value >1 means that the drainage condition is better than that existed in the AASHO site. AASHTO recommends the selection of Cd on the basis of the drainage quality of that specific

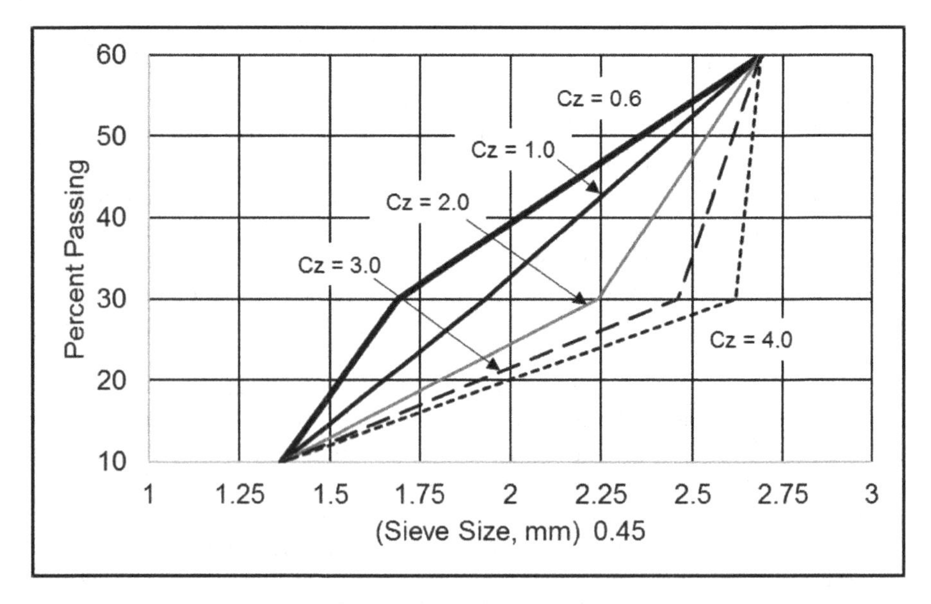

FIGURE 7.16 Possible variation in Cz for the same Cu of 4.728 (permeable base gradation from New Jersey DOT, after Mallela et al., 2000), for fixed D_{60} and D_{10}, but variable D_{30}.

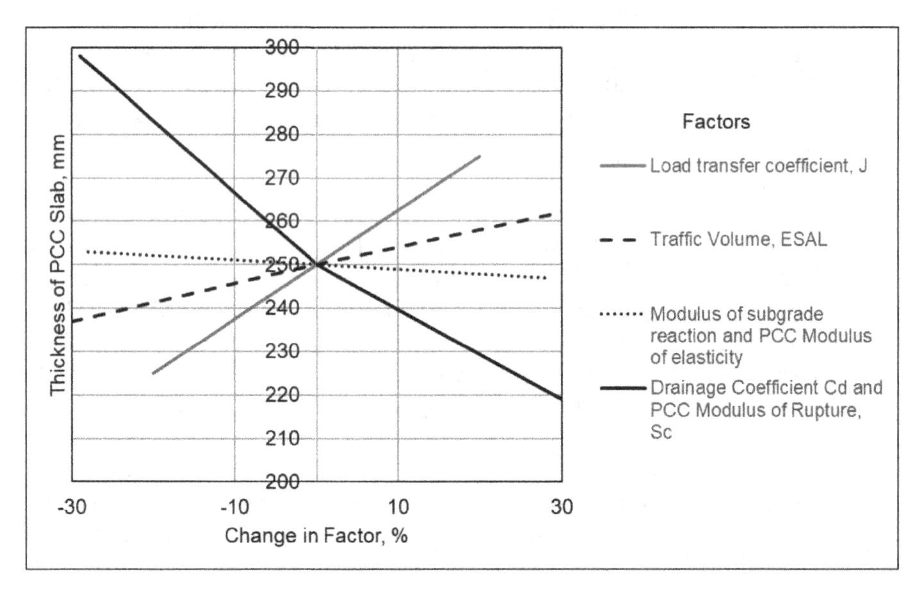

FIGURE 7.17 Sensitivity of PCC thickness to various factors (recreated from Mallela et al., 2000).

TABLE 7.7
Recommended Unstabilized Aggregate Properties

Aggregate Properties	Gradation Type					
	AASHTO #57	AASHTO #67	New Jersey	Pennsylvania	Minnesota	Iowa
D_{10}, mm	6.01	5.78	1.91	2.34	1.14	0.90
D_{30}, mm	10.19	8.53	3.30	5.62	4.70	3.14
D_{60}, mm	15.43	12.37	9.01	14.80	11.74	9.82
C_u	2.57	2.14	4.72	6.31	10.26	10.96
C_z	1.12	1.02	0.63	0.91	1.64	1.12

Source: Mallela et al., 2000.

layer and the percentage of time that the pavement is exposed to saturation conditions. Relevant factors such as the subgrade permeability, freeze thaw, geotextile and separator layer, drainage relief (outlet pipe system or daylighting) are not considered and are for unstabilized layers only.

The method has since been updated in the 1998 supplement in which a better defined Cd is used in the mechanistic based faulting model.

Because of shortcomings the authors have recommended the process for the estimation of the Cd as outlined in Table 7.8.

Mallela et al. (2000) caution that continuous maintenance of the edgedrain should be done to make the permeable base system effective – without which the alternative is not feasible.

The separator layer is different from the filter layer such that the filter layer allows the passage of water, but the separator layer forms an impermeable barrier between the dissimilar materials and provides support to the construction traffic and materials. Treated or untreated dense graded aggregate layers are recommended as the separator layer. The severity is judged by the amount of annual precipitation, design traffic, subgrade strength (CBR) and fines content (passing the 0.075 mm sieve) in the subgrade (Table 7.9). For the condition with firm foundations a granular layer can be substituted with a geotextile layer. For high severity projects both aggregates and geotextiles are recommended. The geotextile should be placed on top of the aggregate layer. Criteria for design of aggregate separator layer are as follows:

1. Separation and uniformity at the base-subgrade interface and separator-base interface that prevents the intermixing of dissimilar layers.
2. Allowable maximum percentage of fines <12% passing the 0.075 mm sieve and Cu minimum >20 – this ensures a low amount of pumpable fines and adequate stability. The thickness of this layer ranges from 150–1300 mm and is treated as a dense graded aggregate layer in the structural design process.

Criteria for the design of the geotextile layer includes soil retention, permeability, clogging, survivability and endurance. Typical successful geotextile is nonwoven (needle punched, polypropylene, stapled fiber) geotextile with a weight/area of 350 gm/m^2 and it is assumed to have no contribution of the structural design.

The edgedrain trench performs the following function:

1. Provide a hydraulic conduit to water from the permeable base to the edgedrain.
2. Provide support to the shoulder.
3. Hold the edgedrain in place without crushing.

The material used to backfill the trench should have at least the same permeability as that of the base material and be able to withstand shoulder traffic without cracking or settling. The recommendation is to stabilize the trench material with either cement or asphalt especially when full depth asphalt shoulders are not used. Typically, the trenches are with a width of 300 mm.

The cost of providing a JPCP with permeable base was reported to be between 114–124% of a pavement without it. The range is due to the use or nonuse of

TABLE 7.8
Guidelines for Determination of Drainage Coefficient

Step 1: Determine the Environment of the Project Location (Dry Non-freeze, Dry Freeze, Wet Non-Freeze, Wet-Freeze)

Dry/wet: precipitation ≤ 508 mm/year/ precipitation >508 mm/year

Freeze/no-freeze: *freezing index >83°C days/ freezing index ≤ 83°C days*

Step 2: Determine the Quality of the Pavement Base Drainage (Excellent, Good, Fair)

Excellent: <2 hours; Good: 2–24 hours; Fair: 24–168 hours

Step 3. Determine the Effectiveness of the Support Features

Separator Layer	**Effectiveness**	
	Edgedrain/Ditch	**Daylighted/Ditch**
Dense-graded aggregate base (DGAB)	Good	Fair
Geotextile	Fair	Poor
DGAB + Geotextile	Good	Fair

Step 4: Determine the pavement drainage quality

Support Features Effectiveness	**Quality of pavement drainage**		
	Excellent	**Good**	**Fair**
Poor	Fair	Fair	Fair
Fair	Good	Good–fair	Fair
Good	Excellent	Good	Fair

Step 1: Determine the Environment of the Project Location (Dry Non-freeze, Dry Freeze, Wet Non-Freeze, Wet-Freeze)

Step 5: Determine the drainage coefficient

Drainage Quality	Environment			
	Dry Non-freeze	Dry Freeze	Wet Non-freeze	Wet freeze
Very poor	1.00–0.90	0.90–0.80	0.80–0.70	0.70
Poor	1.10–1.00	1.00–0.90	0.90–0.80	0.80
Fair	1.15–1.10	1.10–1.00	1.00–0.90	0.90
Good	1.20–1.15	1.15–1.10	1.10–1.00	1.00
Excellent	1.25–1.20	1.20–1.15	1.15–1.10	1.10

Notes:

1. Determine if the pavement is located in a dry or wet climate or a frozen or nonfrozen location based on SHRP criteria.
   Wet climate = Precipitation >508 mm/year
   Dry climate = precipitation <508 mm/year
   Frozen = Freezing index >83°C days
   Nonfrozen = Freezing index ≤83°C days
2. Determine the quality of permeable base drainage based on the time required to drain the base.
3. Determine the effectiveness of the drainage system based on the type of support features available (separator layer, outlets, and ditches) and their respective efficiencies from Step 3.
4. Based on information from Steps 2 and 3, determine the quality of drainage of the pavement from Step 4.
5. Determine drainage coefficient, C_d, from Step 5. The C_d values in the table were adapted from the listed references based on researcher experience.

Source: Recreated after Mallela et al., 2000.

TABLE 7.9
Guidelines for Evaluating Project Considerations and Selecting Separator Layer

Severity of Project in Terms of Drainage	Existing Conditions					Recommendations for Separator Layer
	Rainfall	Subgrade			Traffic	
		Passing 0.075 mm Sieve	Pumping Potential	CBR		
High	>1 m/yr	>15	High	<6	>2*10^6 for asphalt and 3*10^6 for PCC pavement	Combination of aggregate and geotextile separator layer or a stabilized separator layer, which is not recommended for unstabilized permeable base
Moderate	0.65–1 m/yr	Moderate	Low	6–10	<2*10^6 for asphalt and 3*10^6 for PCC pavement	Aggregate separator layer
Low	0.5–0.65 m/yr	–	Low, high permeability >3m/day	>10	<2*10^6 for asphalt and 3*10^6 for PCC pavement	Geotextile or aggregate separator layer

Note: Nonwoven geotextile, 0.030 kg/m3; aggregate separator layer – dense graded base, CBR ≥50.

Source: Recreated after Mallela et al., 2000.

stabilization of the layers. The other costs include edgedrain maintenance of the entire life cycle of the pavement. The economics can be considered by comparing the distress with those of a conventional pavement. Mallela et al. (2008) show an example which indicates that faulting can be reduced by placing a drainage system. The authors recommend a permeable base system for pavements with moderate to high design traffic >4.5 million ESALs in the design lane.

7.14 RISK ANALYSIS FOR PAVEMENT SUBSURFACE DRAINAGE SYSTEM DESIGN

Kalore et al. (2019) provides a framework (Figure 7.18) for the design of the permeable layer with respect to risk analysis. The framework is based on the capacity-demand model and takes into consideration the basic governing equation of flow that is Darcy's law. The demand on the system is considered as the required permeability, which is based on the total inflow into the pavement system and the geometric-section properties. The discharge capacity of the layer is considered as the permeability of layer. The demand and capacity are evaluated in the light of uncertainty and stated in terms of the probability of failure and economic risk. They demonstrate that the associated risks are relatively very high for dense-graded mixes and multi-lane highways, from the drainage point of view. The risk can be reduced by controlling the factors that affect the design. The factors that can be controlled are the permeable layer thickness and the flow length, for a selected aggregate-gradation and given geometry. A sensitivity analysis for the factor of safety (FS) was carried out and the results for a given amount of moisture inflow show the influential parameters (in decreasing order) to be permeability at full saturation (k_s), thickness of the permeable layer (t_t), cross slope (S_x), one-way width of road (w). (Figure 7.19). Their results show a considerable decrease in risk with an increase in the coefficient of permeability (Figure 7.20). The requirement of coarser base layer gradation may lead to poor performance due to a lack of enough stability, which can be resolved by treating and stabilizing it with cement and asphalt. An economical and safe design can be obtained with the use of optimum permeable layer thickness and coarser base layer gradation with stabilization.

7.15 DETERMINATION OF PAVEMENT PERMEABILITY

The Air Permeameter (Figure 7.21, White et al., 2014) can be used to estimate the in-place permeability of granular materials.

The device consists of a contact ring, differential pressure gages, pressure orifices and a programmable digital display. The weight of the device (18 kg) is utilized for maintaining effective sealing to the ground during testing. Compressed gas is used as the permeant for the measurement of flow rate. The test is based on the theoretical relationship between saturated hydraulic conductivity (Ksat) and gas pressure and flow rate measurements, details of which are provided in White et al. (2014).

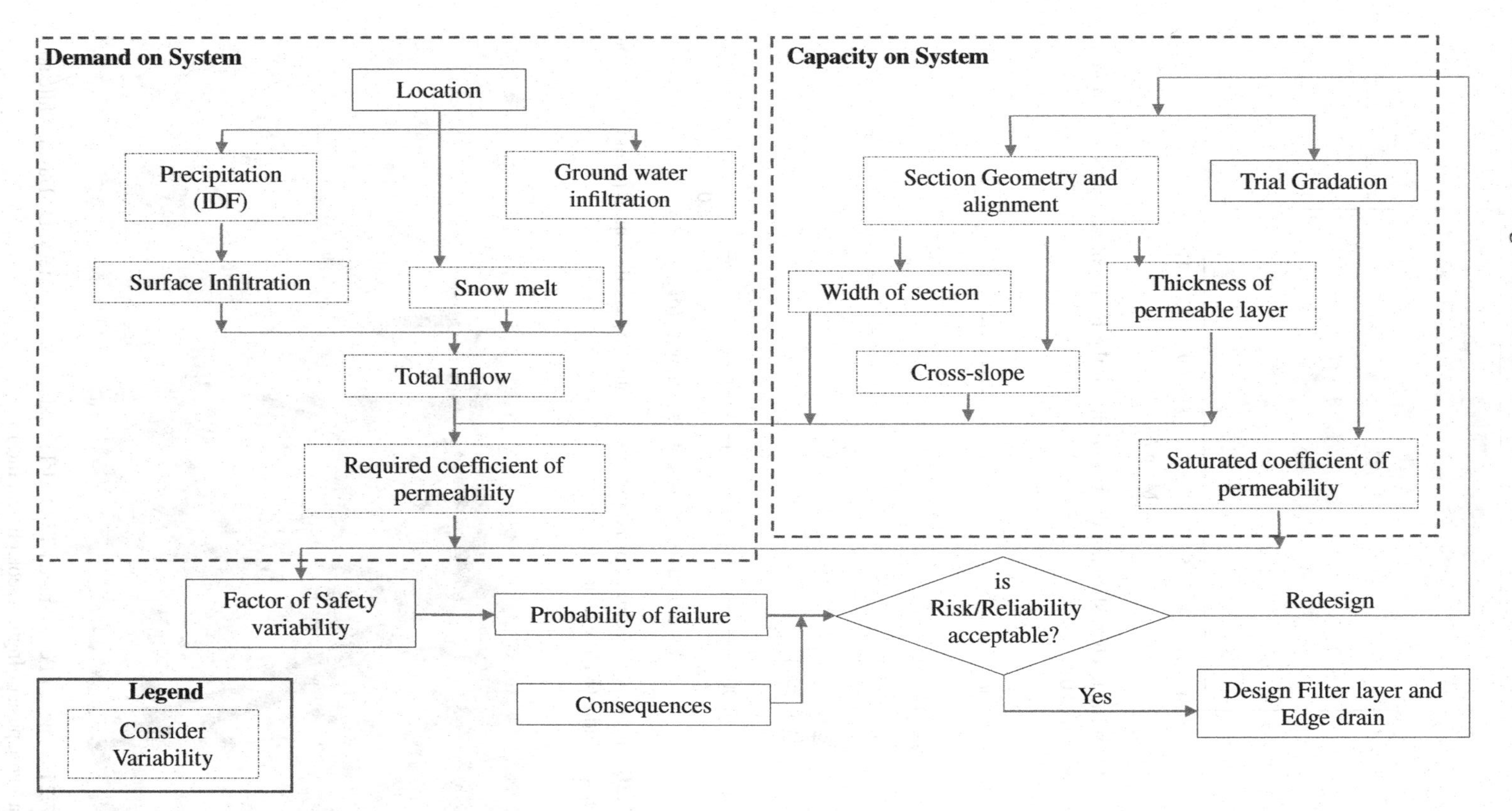

FIGURE 7.18 Framework for risk-based design of the pavement subsurface drainage system (Kalore et al., 2019).

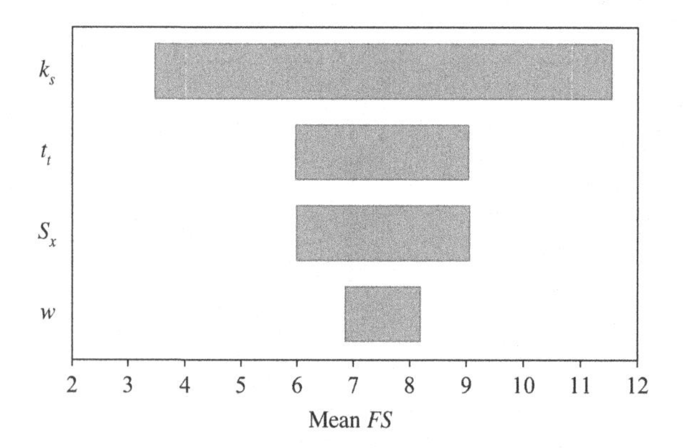

FIGURE 7.19 Tornado chart showing the sensitivity of the Factor of safety to the different factors (Kalore et al., 2019).

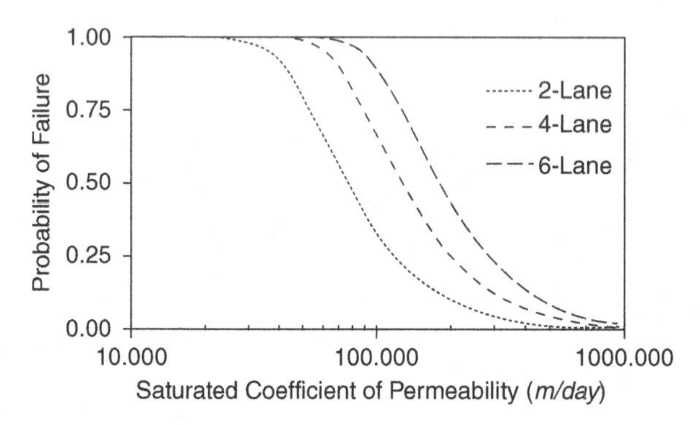

FIGURE 7.20 Plots of probability of failure versus saturated coefficient of permeability for Mumbai region, India (Kalore et al., 2019).

FIGURE 7.21 The Air Permeameter Test (APT) equipment. (Courtesy: Dr. David White and Dr. Pavana Vennapusa of Ingios Geotechnics, Inc.)

REFERENCES

Ariza, P. and Birgisson, B. 2002. Evaluation of Water Flow through Pavement Systems. University of Florida.

Babu, G.L.S. and Srivastava, A. 2014. Analytical Solutions for Granular Filters. Proceedings of Indian Geotechnical Conference, December 18–20, Kakinada Chapter.

Bahador, M., Evans, T.M. and Gabr, M.A. 2013. Modeling effect of geocomposite drainage layers on moisture distribution and plastic deformation of road sections. *Journal of Geotechnical and Geoenvironmental Engineering*, 139: 1407–1418.

Birgisson, B. and Roberson, R. 2000. Drainage of Pavement Base Material: Design and Construction Issues. Transportation Research Record 938, Transportation Research Board, National Research Council. Washington, DC.

Brooks, R.H. and Corey, A.T. 1964. Hydraulic properties of porous media. Hydrol. Pap. 3. Colo. State Univ., Fort Collins, CO, 27 pp.

Burdine, N.T. 1953. Relative Permeability Calculations from Pore Size Distribution Data. Trans. AIME, Vol. 198, pp. 71–78.

Christopher, B.R. and McGuffey, V.C. 1997. Pavement Subsurface Drainage Systems. Synthesis of Highway Practice 239, Transportation Research Board, NCHRP, National Research Council, Washington, DC.

Daleiden, J.F. 1998. Video Inspection of Highway Edgedrains. Federal Highway Administration Report FHWA-SA-98-044.

Federal Highway Administration (FHWA). 1992. Drainable Pavement Systems. Participant Notebook. Office of Pavement Technology, FHWA Report FHWA-SA-92-008.

Federal Highway Administration (FHWA). 1998. Geosynthetic Design and Construction Guidelines. Participants Notebook. FHWA Report HI-95-038, Revised 1998.

Federal Highway Administration (FHWA). 2002. Construction of Pavement Subsurface Drainage Systems (Reference Manual). Office of Pavement Technology, FHWA.

Federal Highway Administration (FHWA). 2008. Pavement Subsurface Drainage Design. NHI Course 131026 Reference Manual. Report FHWA-NHI-08-030.

Fredlund, D.G. and Xing, A. 1994. Equations for the soil-water characteristic curve. *Canadian Geotechnical Journal*, 31: 521–532.

Gardner, W.R. 1958. Some steady state solutions of unsaturated moisture flow equations with applications to evaporation from a water table. *Soil Science*, 85 (4): 228–232.

Green, R.E. and Corey, J.C. 1971. Calculation of hydraulic conductivity: a further evaluation of some predictive methods. *Soil Science Society of America Proceedings*, 35: 3–8.

Harvey, J.T., Tsai, B.W., Long, F. and Hung, D. 1999. CAL/APT Program – Asphalt Treated Permeable Base (ATPB). Federal Highway Administration. Report FHWA/CA/OR – 99/09.

Kalore, S.A., Babu, G.L.S. and Mallick, R.B. 2019. Risk Analysis of Permeable layer in Pavement Subsurface Drainage System. *ASCE's Journal of Transportation Engineering, Part B: Pavements* (in press).

Lambe, R. and Whitman, R. 1969. *Soil Mechanics*. New York: Wiley.

Li, C., Ashlock, J., White, D. and Vennapusa, P. 2017. Permeability and stiffness assessment of paved and unpaved roads with geocomposite drainage layers. *Applied Science*, 7: 718. doi:10.3390/app7070718.

Lovering, W. and Cedergren, H. 1962. *Structural Section Drainage. Proc. First International Conference on the Structural Design of Asphalt Pavements*. Ann Arbor, MI: University of Michigan.

Mallela, J., Titus-Glover, L. and Darter, M. 2000. Considerations for Providing Subsurface Drainage in Jointed Concrete Pavements. Transportation research Record 1709, Transportation research Board, Washington, DC.

Moulton, L.K. 1980. Highway Subsurface Design. Federal Highway Administration Report FHWA-TS-80–224.

Mualen, Y. 1976. A new model for predicting the hydraulic conductivity of unsaturated porous media. *Water Resources Research*, 12 (3): 503–522.

NRCS. 1994. Soil engineering. National engineering handbook. National Resources Conservation Services (NRCS), U.S. Department of Agriculture, Washington, DC. Chapter 26, Part 633.

Richards, B.G. 1974. Behavior of unsaturated soils. In *Soil Mechanics-New Horizons*, Ch. 4. New York: American Elsevier Publishing Company Inc.

Roberson, R. and Birgisson, B. 1998. "Evaluation of Water Flow Through Pavement Systems." Proceedings. *International Symposium on Subdrainage in Roadway Pavements and Subgrades*, pp. 295–302.

Srivastava, A. and Babu, G.L.S. 2011. Analytical solutions for protective filters based on soil-retention and permeability criteria with respect to the phenomenon of soil boiling. *Canadian Geotechnical Journal*, 48 (6): 956–969.

Stormont, J.C., Henry, K. and Robertson, R. 2006. Geocomposite Capillary Barrier Drain (GCBD) for limiting moisture changes in pavements: Product application. www.researchgate.net/publication/241868376.

Stormont, J.C., Pease, R.E., Henry, K., Barna, L. and Solano, D. 2009. Geocomposite Capillary Barrier Drain for Limiting Moisture Changes in Pavements: Product Application Final Report for Highway IDEA Project 113. Transportation Research Board, Washington, DC.

van Genuchten, M. Th. 1980. A closed-form equation for predicting the hydraulic conductivity of unsaturated soils. *Soil Science of America*, 44: 892–898.

White, D.J., Vennapusa, P. and Zhao, L. 2014. Verification and repeatability analysis for the in situ air permeameter test. *Geotechnical Testing Journal*, 37 (2), 1–12. doi:10.1520/GTJ20130111. ISSN 0149-6115.

8 Interlayers

8.1 GEOTEXTILES

The performance of a pavement structure can be improved by using materials or a combination of materials between the pavement layers. These materials or combinations of materials are termed as interlayers (Figure 8.1).

The main function of the interlayer includes resistance to moisture, and mitigation of reflection cracking. They also provide reinforcement to the asphalt layer so that the pavement can withstand traffic loading and thereby extend the life of the pavement structure. The performance of the interlayer will depend on the strength of the interlayer and the asphalt layer. The strength of the interlayer must be more than that of the asphalt layer so that it can act as a stress-relieving layer. Interlayers can be broadly classified as bituminous-based materials (such as chip seals, slurry, asphalt rubber membrane) or geosynthetics (such as paving fabrics, paving mats, composite grids, peel and stick products). Two examples of commercially available products are described below.

8.1.1 RoadDrain™

RoaDrain™ Roadway Drainage System (Figure 8.2) is a commercially available synthetic subsurface drainage layer that can replace conventional base layers. It is an interlayer, which is composed of nonwoven geotextile filters laminated to the top and bottom of a tri-planar geonet core. The geonet core resists deformation by maintaining the void structure with high compressive strength.

The geotextile material helps in separation and filtration of fines and other materials. The main advantages of RoadDrain includes:

- The flow rate is increased by 500%. It quickly removes water and thereby decreasing overall drainage time.
- 50% water is drained out from a pavement structure within two hours.
- The material cost is lowered by reducing the requirement of processed structural fill quantities.

FIGURE 8.1 Synthetic RoaDrain drainage system. (Source: www.tensarcorp.com/Applications/Roadway-Drainage, used with permission.)

- Construction cost is saved and construction schedules are accelerated with easy and faster installation than a conventional base layer.
- Quick removal of water from pavement structures in a high water table environment and act as a capillary break that limits the capillary action of ground water into the base layer.
- Provides separation of the structural base course from the subgrade conditions.
- Provides sufficient compressive stiffness to support traffic loading.

Several case studies of successful implementation of this system are summarized in Figure 8.3.

8.1.2 GlasPave®

Another commercial product, the GlasPave® Paving Mat is a combination of fiberglass mesh embedded in high performance polyester mats. It is suitable for all asphalt pavement surfaces. It has a geosynthetics, non-woven polyester matrix which help the asphalt binder to penetrate and fill the voids thereby reducing moisture ingress into the underlying layers. It also promotes strong bonding with different types of asphalt tack coats. Various advantages of GlasPave® Paving Mat include:

- Provides high tensile strength at low strain and it delays reflective cracking. This helps in fewer maintenance cost.
- Improves the service life of the road by limiting the infiltration of moisture into the pavements structure.
- A high temperature fiberglass matrix, which has good thermal stability and will not shrink as opposed to polypropylene fabrics, which will change dimensions when hot asphalt comes in contact.
- Are durable and less prone to damage during on-site installation.
- The stiff fiberglass aids in trouble-free milling during asphalt layer rehabilitation.

Location: Bodega Highway, Sonoma County, CA; Road was affected by freezing of water that comes up from lower layers. RoaDrain layer was placed between the aggregate base and the silty subgrade soil as a separation layer. It provided a drainage path and strength to the pavement section.

Location: US Route 1, Maine; failure of road due to poor subgrade drainage, migration of fines and failure of aggregate base; unavailability of sufficient amount of low cost aggregate that are required for the design thickness; RoaDrain™ was installed between the base layer and the silty subgrade. It provided a drainage path and reduced the time to drain from months to less than a day. It also acted as a separation layer between subgrade and base layer, and the thickness of base layer was reduced from 24 to 12 inch. while still maintaining service life of the roadway.

Location: Southwest Parway Street, Austin, Texas; ground water table fluctuation caused ingress of water into the base course, which, upon satration caused failure of the pavement; RoaDrain™ was provided under the base course as a drainage conduit to channel the groundwater to a collection system.

FIGURE 8.2 Some successful use of the RoaDrain™ system. (Source: www.tensarcorp.com/Applications/Roadway-Drainage, used with permission.)

Asphalt pavements which have developed fatigue cracking require complete sealing of the cracks to avoid intrusion of rainwater which would cause further deterioration of the pavement structure. Providing a composite (combination) of non-woven geotextile (GT) paving fabric (which is continuous and has polyester and glass fibers) and a glass fiber grid reinforcement is very helpful (Kandhal and Veeraragavan, 2018). Application should consist of a heavy tack coat of paving bitumen binder (not

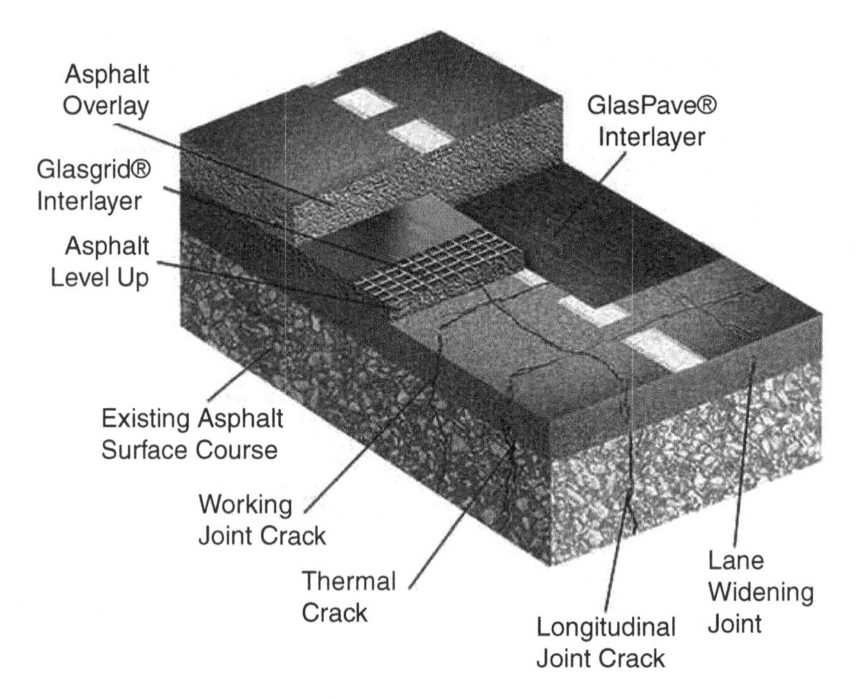

FIGURE 8.3 Schematic showing the placement of interlayers. (Source: www.tensarcorp.com/News-Events/News/What-are-interlayers, used with permission.)

emulsion) at a rate specified by the manufacturer of the paving fabric. The nonwoven geotextile together with a heavy tack coat would ensure "complete sealing" of the existing pavement. The glass fiber geogrid would provide "reinforcement" to the overlying asphalt overlay to resist tensile stresses at its bottom. It would also prevent lateral movement of the overlay asphalt mix.

REFERENCE

Kandhal, P.S. and Veeraragavan, A. 2018. Investigation of premature distresses on typical national highway project. *Journal of the Indian Roads Congress*, 79 (2): April–June 2018.

9 Performance and Need for Maintenance of Drainage Structures

9.1 SURVEY OF STATES

A survey of subsurface drainage systems in Iowa by Ceylan et al. (2013) found that about 20% of the JPCP drainage systems and 10% of the HMA systems were damaged, and most of the blockages are due to tufa, sediment and soil. For the JPCP pavements the primary mode of blockage was by tufa, whereas for the HMA pavements it was by soil deposits. Tufa is a variety of limestone formed when carbonate minerals precipitate out of ambient temperature water.

They observed that the use of Recycled Portland Cement Concrete (RPCC) as subbase material has led to the formation of tufa through precipitation of calcium carbonate. This can potentially be eliminated by using virgin aggregates and reduced by using RPCC/virgin aggregate blend. The potential of deposits of fine aggregates and blockages (Steffes, 1999) can be reduced by washing of the RPCC before its use. The authors noticed that the use of rodent screen (mesh or gate) had the potential to cause blockage and recommended against their use in the absence of any rodent. Pavement shoulder distresses such as drop or cracking adjacent to the blocked outlet areas have been observed.

The use of headwall as outlet protection system for the drainage outlet systems is recommended. It has been noted that if the drainage system is not designed or constructed or maintained properly then it can cause a significant amount of problems and lead to premature distresses in the pavement (Christopher, 2000). Most failures of drainage systems have been due to either poor construction or poor inspection/maintenance practices. Failure of edgedrains can occur if the water from the drainage layer is blocked due to a missing outlet, plugged outlets, crushed outlets, clogged filters or clogged drains (TRB 1997, 2000).

A survey of the results of a video inspection of the pavement drainage systems (FHWA HR-317, Daleiden, 1999) concluded that of the drainage systems that could be inspected (66%), half were working properly and the remaining half were non-functional because of various reasons. They presented a guide for video inspection of the edgedrain and acceptance.

The problems and issues about subsurface drainage systems have been listed as follows (Hall and Crovetti, 2007): Inadequate design, improper construction, inadequate maintenance, use in locations where not needed (low rainfall areas), concerns about construction difficulties, need to conduct frequent maintenance of edgedrains and inadequate evidence of performance benefits that would justify the installation and maintenance cost for the drainage system.

The authors report that in some instances the presence of subsurface drains have helped in correcting faulting problems.

The overall impression has been that subsurface drainage systems help to prolong the service life of pavements. If pavements are already experiencing distress due to other factors then the lack of subsurface drainage systems or non-functional drains tend to accelerate the process through freeze-thaw problems and PCC joint problems. The principal causes of the blocking of drains have been identified as a formation of tufa (where RPCC is used as subbase), vegetation formation and deposits of silts and sediments in the drainage outlets. Shoveling near the outlets can be an effective treatment for the removal of blockages in many cases.

Hall (2002) reported that day-lighted permeable bases in HMA pavements with unbound aggregate bases were more effective in reducing fatigue cracking than those with edgedrains, permeable bases with clogged edgedrain outlets led to increase in fatigue cracking and rutting, and that permeable base courses were effective in reducing joint faulting in un-doweled JPCP, and in reducing D-cracking in PCC pavements. These benefits could lead to an extension of the pavement lives and hence delaying rehabilitation activities. Based on life cycle cost analyses the author mentioned that the use of subsurface drainage should be compared against other options such as using thicker layers of pavements. Hall and Correa (2003) reported on an extensive study on the use of LTPP data to evaluate the effects of subsurface drainage. They concluded that HMA pavements with subsurface drainage systems with asphalt treated permeable base courses performed better in terms of IRI and cracking than those with undrained dense-graded aggregate bases, but were inferior to those with undrained dense graded asphalt-treated base courses. For PCC pavements, those with drained permeable asphalt-treated bases showed better performance in terms of IRI, transverse cracking and longitudinal cracking compared to pavements with undrained dense-graded aggregate bases as well as undrained lean concrete bases. Hall and Crovetti (2007) conducted a study to refine the relationships between subsurface drainage and pavement performance that were derived earlier, on the basis of newer LTPP data, and extensive field testing for permeability (inflow and outflow from permeable base using flow meter) as well as analysis of FWD data. Even though the regression statistics did not show significant proof of beneficial effects of the subsurface drainage system on the performance of HMA and PCC pavements, the authors did recommend the use of subsurface drainage systems for projects in wet climates and poorly draining soils, and specifically for designs that would be vulnerable to moisture distress such as thin asphalt or thick concrete pavements on untreated aggregate base layers.

Hassan et al. (1996) documented the long-term research effort by Indiana DOT for the evaluation of subsurface drainage systems through the use of TDR and neutron probe for measuring moisture content and thermocouples and resistivity probes for measuring pavement temperature and depth of frost penetration, respectively.

Zubair et al. (1993) conducted an extensive investigation of subsurface drainage systems by using instrumentation such as pressure transducer, moisture blocks, thermistor probe, rain gauge, tipping bucket and flow meter, and presented a guideline for DOTs to monitor the condition of subsurface drainage systems.

A finite difference program was developed by Espinoza (1993) (PURDRAIN) to analyze water flow and drainage in pavements under unsaturated conditions.

The Indiana Department of Transportation (INDOT) discontinued the use of geocomposite drains and replaced them with Group K pipes (for edgedrains) in the mid-1990s, adopted a routine inspection and maintenance program and made it mandatory for all construction projects to inspect and repair edgedrains as necessary by the contractor. Based on an evaluation of edgedrains and center drains in Minnesota, Canelon and Neiber (2009) concluded that there were no significant differences between the drainage volumes of centerline drains at 2 ft and 4 ft centerline depths, more moisture was present within the edgedrains than the centerline drains, and carbonate sands have the potential to lead to deposits in the drainage system. They recommended edgedrains over centerline drains but mentioned that centerline drains could be effective in pavements with a permeable subgrade.

Arika et al. (2009) developed a manual on evaluation of need for subsurface drainage, selection of type and design of drainage system, guidelines for construction of subsurface drainage, their proper maintenance and economic analysis.

A study by Bhattacharya et al. (2009) on subsurface drainage systems in PCC pavements in California revealed that less than 30% of the systems which were in areas of high rainfall, were functioning properly. The problem with the remaining sites were clogging of the drainage system underneath the road and in the shoulder, which were aggravated by the absence of headwalls. The authors concluded that larger diameter drain pipes, deep trenches and treated permeable bases provided better performance over those with retrofitted edgedrains with slotted pipes, primarily because of the shallow location of the retrofit edgedrain trenches which prevented them from the effective collection of water from the PCC and the base layers, and faulty construction which resulted in the placement of several edgedrains on a high side of the cross slope. They recommended that edgedrains be used only in critical drainage areas with a commitment for maintenance, and after a careful review of rainfall and permeability of the natural soil. The geotextile filters should be placed along the side of the shoulder and trench bottom and should be selected after proper consideration of the existing soil. They also recommended larger (4 in) diameter slotted pipes for ease of video inspection and dual outlets for easy maintenance. Their overall conclusion was that the presence of edgedrains will not provide additional benefits for performance of PCC pavements with proper dowel bars and tie bars if day-lighted permeable bases and HMA layers (possibly as base or subbase course) are already present. Baumgardner (2002) noted that the maintenance of subsurface drainage systems can be facilitated greatly by headwalls, reference markers, signs on fences, reflector discs on shoulders or painted areas on the shoulder. Large headwalls are preferred because it helps in locating the drainage outlet pipes, avoids roadside vegetation, reduces erosion at the drainage outlet and helps avoid damaging the outlet pipes during construction and mowing operations.

The subject of using Recycled Concrete Aggregate (RCA) in unbound subbase and base in concrete pavements has been studied by a number of researchers (Gupta and Kneller, 1993; Snyder and Bruinsma, 1996; Steffes, 1999; White et al., 2008;

Phan, 2010). The conclusions are that RCA leads to precipitation which is directly related to the quantity of fines (percentage passing the 4.75 mm sieve), the potential of precipitation can be reduced by washing of the RCA, the precipitation reduces the permittivity of the geotextile filter, rodent guard screens used in outlet pipes should be made of corrosion resistant material (corrosion is caused by the precipitates from the RCA), larger diameter pipes either unwrapped or wrapped with geotextile fabric with high initial permittivity are desirable, and the calcium ion concentration test (recommended by MI and MN DOT) may be a good test to evaluate the precipitate potential of the RCA.

Ceylan et al. (2013) carried out a detailed forensic investigation of subsurface drainage systems in Iowa. They included both JPCP and HMA pavements, with high (2,000–14,000) and relatively low (1,000–1,750) AADTT, respectively. The JPCP pavement age ranged from 10–20 years, while the HMA pavements were 5–10 years old. A range of pavement thicknesses (10–13 inch for PCC slab and 10–16 inch for HMA layer) and PCI (60–100 for PCC, 70–80 for HMA) were included. Eighty percent of the JPCP pavements used RPCC as base materials. The field investigation consisted of the evaluation of subsurface drainage outlet conditions and a visual distress survey of areas adjacent to the outlet pipes. The researchers rated the condition of the outlet pipes in terms of blocked areas. Information was also collected for the specific sites from the Pavement Management Information System. Their evaluation indicated that less than 20% of the JPCP and less than 10% of the HMA drainage outlets were damaged. Regarding blockage, deposition of calcium carbonate (tufa) was noted only in JPCP pavements with RPCC base. The occurrence was less for JPCP pavements with blends of RPCC and virgin aggregates in the base layers. Sediment and soil deposits were other factors that contributed to blocking in general. About 65% of JPCP and 40% of HMA outlets were found to be blocked. Outlets in HMA pavements were primarily found to be blocked by soil. The authors reported that most surface distress on the pavements were found near open subsurface drainage outlet and not near the blocked ones. However, shoulder distress (drop or cracking) was found more near the blocked drainage outlet spots than near the open ones. Regression analysis of PCI (y) with construction year, AADTT, thickness, and blockage rate of outlet (%) showed no significant impact of blockage rates on the PCI.

REFERENCES

Arika, C.N., Canelon, D.J. and Nieber, J.L. 2009. *Subsurface Drainage Manual for Pavements in Minnesota.* Final Report, MN/RC 2009–17. MN: University of Minnesota, Minnesota Department of Transportation.

Baumgardner, R.H. 2002. *Maintenance of Highway Edgedrains.* U.S. Department of Transportation, Federal Highway Administration.

Bhattacharya, B.B., Zola, M.P, Rao, S., Smith, K. and Hannenian, C. 2009. Performance of edge drains in concrete pavements in California. Proceedings of National Conference on Preservation, Repair, and Rehabilitation of Concrete Pavements. St. Louis, Missouri, April 21–24, 2009, pp. 145–158.

Canelon, D.J. and Nieber, J.L. 2009. *Evaluating Roadway Subsurface Drainage Practices*. MN/RC 2009-08. Minnesota: Minnesota Department of Transportation Research Services Section.

Ceylan, H., Gopalakrishnan, K., Kim, S. and Steffes, R.F. 2013. *Evaluating Roadway Subsurface Drainage Practices*. Institute for Transportation Iowa State University.

Christopher, B.R. 2000. *NCHRP Synthesis of Highway Practice 285: Maintenance of Highway Edgedrains*. Washington, DC: Transportation Research Board (TRB) of the National Academies.

Daleiden, J.F. 1999. *Video Inspection of Highway Edgedrain Systems*. Report No. FHWA-SA-98-044, Final Report, Federal Highway Administration (FHWA), McLean, Virginia.

Espinoza, R.D. 1993. *Numerical Analysis of Unsaturated Flow in Porous Media*. PhD Thesis. West Lafayette, IN: Purdue University.

Gupta, J.D. and Kneller, W.A. 1993. *Precipitate Potential of Highway Subbase Aggregates*. Final Report FHWA/OH-94/004, Ohio Department of Transportation.

Hall, K.T. 2002. "Performance of Pavement Subsurface Drainage." *NCHRP Research Results Digest* No. 268. Transportation Research Board (TRB) of the National Academies, Washington, DC.

Hall, K.T. and Correa, C.E. 2003. *Effects of Subsurface Drainage on Performance of Asphalt and Concrete Pavements*. NCHRP Report 499. Transportation Research Board (TRB) of the National Academies, Washington, DC.

Hall, K.T. and Crovetti, J.A. 2007. *Effects of Subsurface Drainage on Pavement Performance: Analysis of the SPS-1 and SPS-2 Field Sections*. NCHRP Report 583. Transportation Research Board (TRB) of the National Academies, Washington, DC.

Hassan, H.F., White, T.D., McDaniel, R. and Andrewski, D. 1996. "Indiana Subdrainage Experience and Application." *Transportation Research Record: Journal of the Transportation Research Board*, No. 1519, pp. 41–50, TRB, National Research Council, Washington, DC.

Phan, T.H. 2010. Laboratory and Field Investigations of Recycled Portland Cement Concrete Subbase Aggregates. Ph.D. Thesis. Iowa State University, Ames, IA.

Snyder, B. and Bruinsma, J.E. 1996. "Review of Studies Concerning Effects of Unbound Crushed Concrete Bases on PCC Pavement Drainage." *Transportation Research Record: Journal of the Transportation Research Board*, No. 1519, pp. 51–57, TRB, National Research Council, Washington, DC.

Steffes, R. 1999. *Laboratory Study of the Leachate from Crushed Portland Cement Concrete Base Material. Materials Laboratory Research Project*. MLR-96-4, Final Report, Iowa Department of Transportation, Ames, IA.

TRB. 1997. *NCHRP Synthesis of Highway Practice 239: Pavement Subsurface Drainage*. Washington, DC: TRB, National Research Council.

TRB. 2000. *NCHRP Synthesis of Highway Practice 285: Maintenance of Highway Edgedrains*. Washington, DC: TRB, National Research Council.

White, D.J., Ceylan, H., Jahren, C.T., Phan, T.H., Kim, S., Gopalakrishnan, K. and Suleiman, M.T. 2008. *Performance evaluation of concrete pavement granular subbase – pavement surface condition evaluation*. Final Report, IHRB Project TR-554, CTRE Project 06-250, Ames, IA.

Zubair, A., White, T.D. and Bourdeau, P.L. 1993. *Pavement Drainage and Pavement Shoulder Joint Evaluation and Rehabilitation*. Joint Highway Research Project. FHWA/IN/JHRP-93/2, Draft Final Report. FHWA, U.S. Department of Transportation.

10 Sustainable Drainage

10.1 POROUS ASPHALT PAVEMENTS (PAP)

Mostly used for walkways, trails, parking lots and low volume roads, porous pavements (FHWA, 2015) (Figure 10.1) offer a process to manage storm water in an environmentally friendly way and is considered to be a multifunctional low impact development (LID) technology (see Figure 10.2, for integrating it with rainwater harvesting).

In porous asphalt pavements (PAP) (Thelen and Howe, 1978; Jackson, 2003; Kandhal, 2011; Kandhal, 2013; Kandhal and Mishra, 2014), the subgrade is left uncompacted to maximize water infiltration. A geotextile fabric is placed between the subgrade and the stone recharge bed above it to prevent migration of fines from the subgrade to the storm recharge bed. The stone recharge or reservoir layer is made up of uniformly graded, clean crushed stone with 40% voids. It serves as a structural layer and temporarily stores water before it infiltrates into the soil below. To stabilize and prepare the top of the reservoir stone layer for paving, a 25 mm (1 inch) thick layer of clean, smaller sized crushed stones is placed on the top (stabilizing or the choker course). On top of this, one or more layers of open graded asphalt mixes with inter connected voids are placed. This open graded mix does not contain any fines and has voids ranging from 16–22%.

A typical cross-section of the porous asphalt pavement system is shown in Figure 10.3. The pavement consists of the following components from top downwards:

- Open-graded, porous asphalt course 50–100 mm (typically 75 mm) thick.
- 12.5 mm nominal size aggregate choking layer 25–50 mm thick (this is placed over the stone bed so as to stabilize it and facilitate asphalt paving over it).
- Clean, uniformly graded 40–75 mm size crushed aggregate compacted layer to act as a water reservoir.
- Non-woven geotextile to separate the soil subgrade and water reservoir course so that soil particles do not migrate from the subgrade into the stone water reservoir course thus choking it. Alternately, a 75 mm thick stone filter course

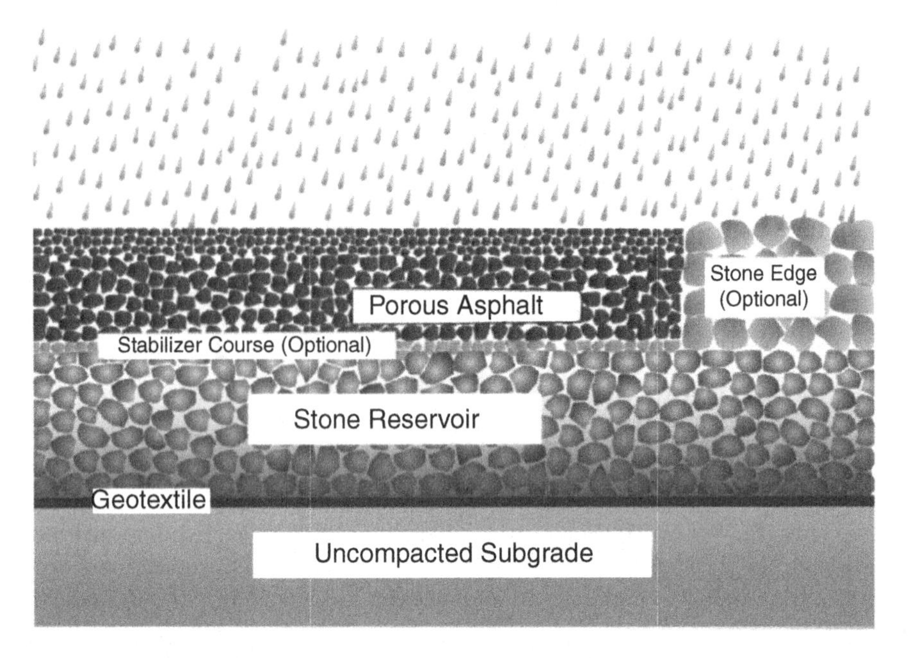

FIGURE 10.1 Concept of porous pavement. (Courtesy, Figure 1. Page 1, Porous Asphalt Pavement with Stone Reservoirs. The document is accessible at the following site of the FHWA: www.fhwa.dot.gov/pavement/asphalt/pubs/hif15009.pdf.)

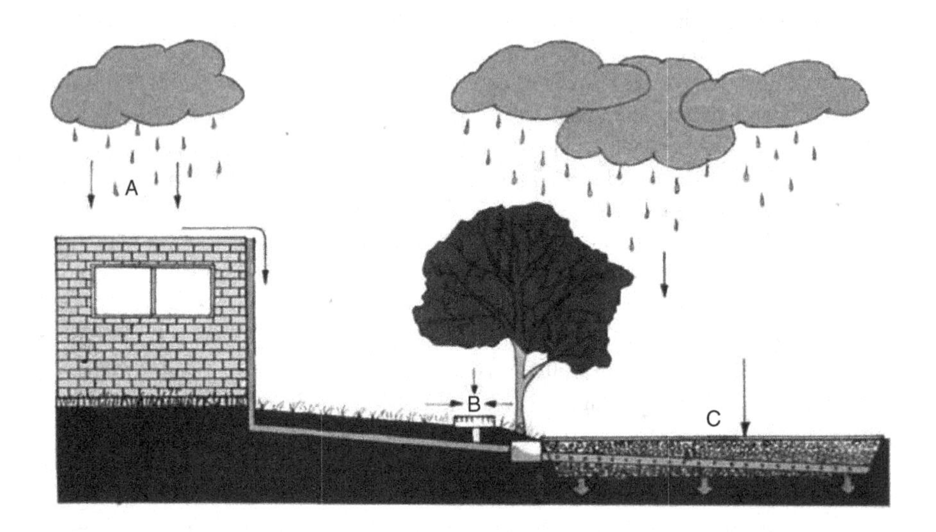

FIGURE 10.2 Porous asphalt pavement integrated with rainwater harvesting. (Courtesy, NAPA.)

FIGURE 10.3 Typical porous asphalt pavement cross section. (Courtesy, Maine DOT.)

consisting of 10–25 mm size aggregate can be provided if good aggregate gradation control can be maintained.

- Uncompacted natural soil subgrade (bed).

A rooftop rainwater harvesting systems of the buildings adjacent to porous parking lots or streets can be integrated into the porous asphalt pavement (as shown earlier in Figure 10.2). The water from the rooftop is carried directly to the stone water reservoir and dispersed there through a series of perforated water pipes. This way, the stone reservoir does not experience any localized flooding. This system also means no soaking well or bore hole which involves considerable cost. In case of streets, water from the rooftop of the buildings on the street can all be diverted to the stone water reservoir course. Another major advantage of this technology is that the water recharging the underground water is pure and free of contaminants.

The benefits of PAP include the aid in storm water management, improvement of water quality, reduction of storm water runoff, reduction of the need for de-icing chemicals in winter, reduction in noise, increase in wet-weather friction and visibility and reduction in storm water temperature before discharge. Porous pavements have been found to mitigate urban heat island effect by reducing stored pavement energy and rapid cooling by evaporation. The major drawbacks of PAP is its structural capacity which currently only allows light traffic. In addition, application in roads and

FIGURE 10.4 View just after rain: parking lot in the background is dense asphalt with water still ponding on it; parking lot in the foreground is porous asphalt. (Courtesy NAPA.)

highways is challenging because of the variability in soil conditions, utilities, fills and slopes. Therefore, their use is mostly restricted to parking lots and low-volume roadways, pedestrian walkways, sidewalks, driveways, bike lanes, shoulders, residential and urban streets.

The dramatic performance of porous asphalt pavements in the US is clearly visible in Figure 10.4 which shows two parking lots just after rain. The one in the background is a conventional dense asphalt parking lot whereas the one in the foreground is a porous asphalt parking lot. Their relative appearance after rain is so very clear.

Figure 10.5 shows a view of a highway in Chandler, Arizona in the US during rain. The left lanes were constructed with porous asphalt and the right lanes were constructed with conventional dense asphalt. After 20 years in service, the porous asphalt on this highway is still functional.

It is absolutely clear that the porous asphalt technology works. Ninety-five percent of the rainwater falling on porous asphalt pavement goes to recharge the ground water. Therefore, its effectiveness in capturing rainwater is very close to paved catchment areas.

10.1.1 Design of PAP

The primary considerations of the design of PAP are:

FIGURE 10.5 Highway in Chandler, Arizona during rain; left lanes are porous asphalt and right lanes are conventional dense asphalt. (Courtesy NAPA.)

1. Site.
2. Hydrology.
3. Traffic loading.

The site should be acceptable, while the thickness of the stone reservoir bed should be adequate for meeting the potential runoff demand from the hydrologic consideration. Finally, the thickness of the surface layer is controlled by the traffic loads. Site considerations include the following:

1. Best for upland soils.
2. Soil infiltration rate of 0.1 to 10 inch/hr is recommended by EPA.
3. There should not be sinkholes.
4. Minimum depth to bedrock or seasonal high water is 2 ft.
5. Infiltration bed should be flat – berms may be constructed to retain water under pavement surface on slope and drains/overflows can be placed at low points.
6. For parking areas, maximum slope of the surface is 5% – if greater, parking areas should be terraced with berms in between.
7. Facilities to convey runoff from nearby impervious areas to infiltration bed; impervious to pervious area should be <5:1 for most conditions and 3:1 for suitable susceptible areas.

Hydrologic considerations for the design of PAP include layer thickness, subgrade permeability and precipitation intensity. Storm water modeling should be done with the SCR/NRCS method and not the Rational method. The PAP is not intended to store and infiltrate maximum precipitation, and overflow provisions should be made to prevent stored water inside PAP from reaching the surface layer. Perforated pipes are placed in the stone reservoirs that are connected to discharge pipes. Alternate paths such as a stone edge or drop inlet could be provided for storm water to reach the stone reservoir, in case the surface becomes plugged.

Structural design is conducted according to AASHTO 1993 for providing areas with truck traffic whereas for those with light car traffic, the design is governed for hydrologic and minimum thickness.

The recommended layer coefficients for different layers of PAP are as follows: Porous asphalt: 0.40–0.42; APTB: 0.30–0.35; Stone recharge bed: 0.10–0.14.

The recommended minimum thickness of PAP are as follows: parking, little or no trucks, 2.5 inch; residential streets with some trucks, 4.0 inch; heavy trucks, 6 inch.

The mix design of porous asphalt is conducted with samples compacted either by 50 gyrations of Superpave gyratory compactor or 35 blows of Marshall hammer. The two critical elements of design are providing adequate air voids for drainage and low draindown for asphalt binder. Generally, fibers and polymers are added to reduce draindown and prevent raveling. Specification of OGFC (Open Graded Friction Course or Porous Friction Course, PFC) can be used for designing the upper layer of PAP (Kandhal, 2002). The main criteria are shown in Table 10.1.

General antistripping agents and/or running moisture susceptibility tests are recommended for porous asphalt.

TABLE 10.1
Mix Design Recommendations for Porous Asphalt

Air voids (AASHTO T 269-11/ASTM D3203M-11)	>16%
Draindown (AASHTO T 305–09/ASTM 6390-11)	≤0.3%
Asphalt content (by weight of total mix)	For 9.5 mm nominal aggregate size porous asphalt mixes, the recommended minimum asphalt content is 5.75% by weight of mix. In rare cases this may not be possible. In these cases the Cantabro test (ASTM D7064M-08) should be run to assure durability. For larger stone mixes, the asphalt content will decrease. The asphalt content should be the highest possible without exceeding draindown requirements.
Maximum aggregate size	The maximum recommend aggregate size for surfaces is 12.5 NMAS. Larger NMAS mixes are recommended for base courses.

Source: Adapted from Hansen, 2008, as mentioned in FHWA, 2015.

10.1.2 Construction and Maintenance

One of the primary requirements is protecting the pavement during construction from runoff from adjacent areas. The surrounding soil should be compacted and temporary storm water control should be arranged during construction. The guidelines include the following:

1. Construct the PAP as late as possible in the construction schedule.
2. Protect the site from excessive heavy equipment on the subgrade to avoid compact subgrade and reduce permeability.
3. Excavate the subgrade soil using equipment with oversize tires or tracks.
4. Immediately upon completion of excavation to fill grade, the fabric filter should be placed with a minimum overlap of 16 inch. A 4 ft excess fabric should be folded over stone bed to protect it from sediments.
5. As required, install drainage pipes.
6. The aggregate of the stone reservoir layer should be placed carefully to avoid damaging the fabric. The material should be stored at the edge of the bed and then placed in layers of 8–12 ft using tracked equipment and compacted with a single pass of a light roller or vibratory roller compactor.
7. If a stabilizer course is used, the appropriate size should be selected so as to interlock with those of the store reservoir layer. The stabilizer course should be placed at a thickness of 1 inch with some of the larger stones for the stone reservoir visible at the surface.
8. Porous asphalt should be placed at a thickness of 1–4 inch preferably using tracked pavers using guidelines for Porous Friction Course or Open Graded Friction Course. It should be compacted with 2–4 passes of 10 ton static roller and traffic should be restricted for at least 24 hours after final rolling.

For maintenance, surface infiltration rates should be inspected annually during rain events and any solids or debris that can lead to clogging should be removed. Finally, the PAP should be vacuumed or power washed 2–4 times every year. In winter, de-icing chemicals can be used for preventing ice/snow accumulation on the surface. Seal coating or cracking sealing is prohibited for porous asphalt. Patching can be done with conventional mixes if the patching area is less than 10% of the pavement area.

10.2 PERVIOUS CONCRETE

Pervious concrete (PC) (Chandrappa and Biligiri, 2016) has been used as a means for reducing storm water runoff. The basic principle of operation of a PC layer is the quick transmission of storm water through an interconnected macro-pore internal structure. This structure is created by selecting appropriate coarse aggregate gradation and cementing materials. The gap graded material is created by single sized coarse aggregates with an optimal amount of cement to coat and bind the aggregates together. A minimum porosity of 15% and a typical range of 15–25% are specified. The water to cement ratio (w/c) is selected at 0.2–0.4, generally lower than that of conventional concrete.

The volume of aggregate ranges from 50–65% with an aggregate to cement ratio of 4:1 to 6:1. The size of aggregates ranges from 19–9.5 mm with smaller aggregates (9.5–2.36 mm) being used occasionally. The physical properties of aggregates must meet the same specifications as those used in ordinary PCC. Generally, PC with granitic aggregates have been found to be more resistant to freeze thaw (F/T) damage compared to those with limestone or river gravel. The relatively thin cement paste in PC allows easier ingress of water into aggregates and hence the properties of the aggregates are very important for the development of appropriate resistance against F/T damage.

PCs are generally produced with ordinary portland cement and an amount that provides for coating of sufficient durability. The use of supplementary cementing materials have been found to be effective up to a threshold (Fu et al., 2014). The coating thickness (6–9 mm) is a significant factor that affects both structural as well as hydrological characteristics of the PC. PCs are designed with zero slump and admixtures are utilized to aid in workability. Admixtures are used to aid in placement and for reducing the evaporation rate, as well as for improving the F/T durability (air entraining agents).

10.2.1 Design of PC

The basic principle of the mix design of PC is to provide an adequate cement coating of the aggregates, which can be achieved by various methods such as the absolute volume method, excess paste theory or on the basis of ratio of paste volume to interparticle voids (Deo and Neithalath, 2011; Nguyen et al., 2014; Yahia and Kabagire, 2014). The typical aggregate density is 1,400–1,800 kg/m^3 with an aggregate to cement ratio range of 4:1 to 12:1.

Typically, the important mechanical properties include compressive strength, flexural strength, fatigue life, abrasion and F/T resistance. It has been seen that the flexural strength of PC remains unaffected by the curing period – 7-day strength is 90% of the 28-day strength. Fatigue life of PC has been found to be low and polymers have been used to improve them. Mix porosity has been found to be a significant factor in dictating fatigue behavior. Generally, PCs with large size aggregates show high compressive strength due to a higher degree of heterogeneity in the pore path of the mixes.

It has been noted that the number of compaction layers may need to be increased in order to reduce the vertical porosity distribution and the potential of failure at the bottom surface where the porosity could be higher. The abrasion resistance of the PC can been evaluated by the Cantabro test or the loaded wheel abrasion test of the surface abrasion test. Fiber and latex have been used to improve the abrasion resistance of PC. ASTM C666 can be used for evaluating the F/T resistance of PC, which can store water in the macropores and hence end up having frozen water under low temperatures.

The properties of the pores in C are crucial in dictating the strength properties. There are two types of properties – Non Transport (NT) and Transport (T) related. NT properties include total volumetric porosity, pore size and distribution. The T properties include effective porosity, pore connectivity and tortuosity. Total porosity is measured by ASTM C 1688, and in reality varies with the depth. Average effective pore size has been reported in a study as 3.4 mm (Kuang et al., 2010).

Strength has been found to decrease by approximately 50% for every 10% increase in porosity (Deo and Neithalath, 2010).

Transport pore properties are actually instrumental in conveying the water from the surface to the bottom of the PC. Total porosity in PC is made up of interconnected, capillary and dead end pores. The effective porosity is the interconnected pore fraction (50–75% of total). An increase in paste content decreases pore connectivity which has also been found to be dependent on aggregate type.

The permeability of PC depends on aggregate size, compaction level, gradation and cement content and is reported to be between 0.1 to 2 cm/s. Chandrappa and Bilgiri (2016) provides an excellent summary of relation between mechanical properties, porosity and permeability (Figure 10.6).

Generally, the strength of the PC is adequate for supporting low volume traffic. The voids get clogged with debris and dust and the infiltration capacity of the PC, and hence the efficiency to reduce storm water runoff is reduced. It has been reported that the top 25 mm of the PC gets clogged. The clogging potential is lower at high- or low-pore size to particle size ratio and the maximum is when the pore size and the clog material sizes are similar.

A typical PC pavement consists of a PC layer, upper optional geotextile layer, base course, geomembrane and the subgrade (Figure 10.7)

In addition, a choker or drainage layer can also be provided. The lower geomembrane can be pervious or impervious depending on the need for water infiltration into the subgrade or jot. Typical distresses with PCs include joint deterioration, surface scaling, raveling and debonding. The structural strength of the PC pavement depends to a large extent on the thickness of the aggregate base course. FWD studies have shown the back calculated modulus values of PCs to be 1/4th that of conventional PCC. PCs have been found to be successful in mitigating the non-point pollutant of runoff by effectively removing the pollutants and hence improving the quality of the infiltrated water in terms of heavy metal, oils, fecal materials and dust. There are 3 mechanisms by which PCs work in reducing pollutants:

1. Physical purification in which most of the suspended solids are absorbed in the curvy paths of the internals structure.
2. Chemical purification in which it (alkaline) releases OH- and CO3- ions in contact with polluted water; the ions react with the pollutants and precipitate them and the pH of the water is also increased, making it neutral from acidic.
3. Biological purification, in which a number of microbes that exist in the pores help in consuming and dissolving the suspended solids.

Illgen et al. (2007) and Illgen (2008) provide some conclusions from a comprehensive field and laboratory infiltration, and modeling study of different types of "permeable" pavements in Germany. They noted that infiltration rates vary both spatially and temporarily. For example, the infiltration rate between the wheel paths were found to be significantly higher than that on the wheel path where clogging may have resulted from tire impact. Also, depending on the type of the pavement (for example, block or

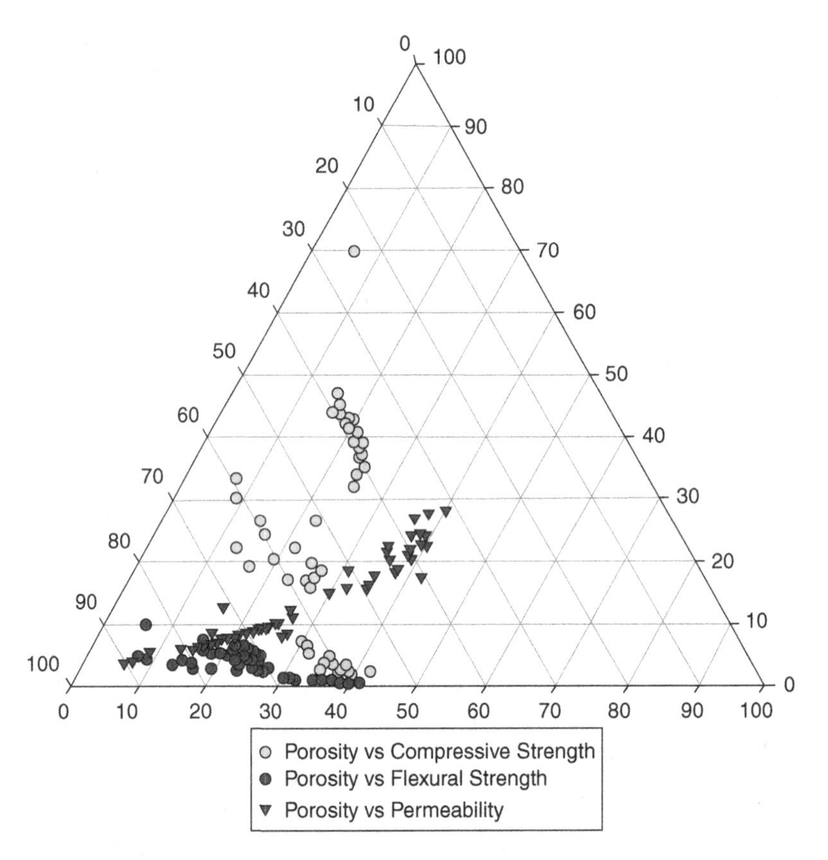

FIGURE 10.6 Interrelationships between different properties of pervious concrete. (With kind permission from Professor Krishna Prapoorna Biligiri.)

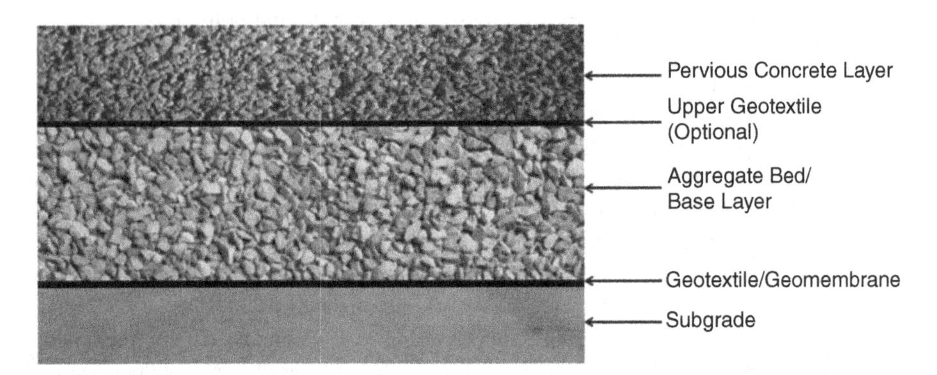

FIGURE 10.7 Typical cross section of pervious concrete. (With kind permission from Professor Krishna Prapoorna Biligiri.)

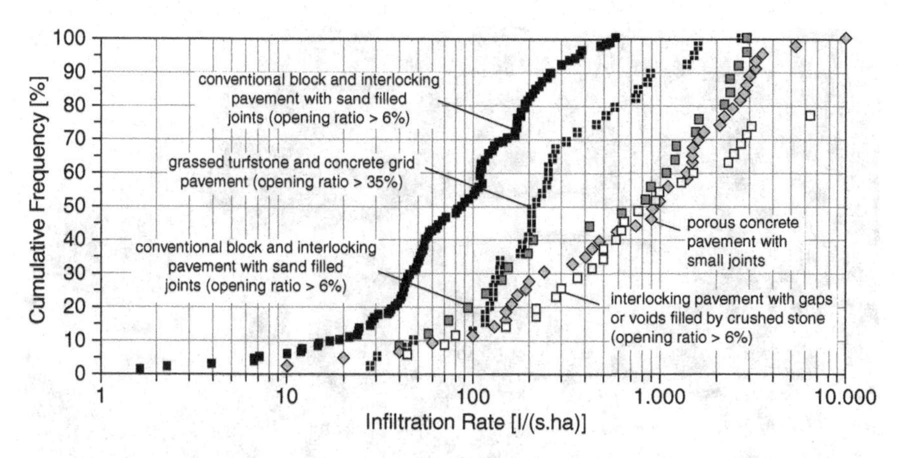

FIGURE 10.8 Cumulative frequency of final infiltration rates of common pavement types. (with kind permission of Prof. Dr.-Ing. Marc Illgen.)

porous concrete pavement), the infiltration rate after 60 minutes was found to be significantly lower than that after 10 minutes of the start of the infiltration. They conclude that infiltration rates vary so widely over a specific site due to many factors such as localized clogging due to construction or impact of tire (see, for example, Figure 10.8, from Illgen, 2008) that it makes more sense to adopt a probabilistic or stochastic approach to this problem rather than specifying or considering a single infiltration rate. They also noted that for a specific pavement the infiltration rate depends significantly on the rain intensity – the infiltration rate increases with an increase in the rain intensity and then levels off to a constant value beyond a specific rain intensity.

An increase in the thickness can lead to an increase in the potential of purification. Studies have reported a removal of 94.3% of phosphorous. Positive experiences have been reported regarding the removal of both heavy metals (industrial area) and fecal matters and phosphorous (sewage).

PCs have been found to lower the life cycle cost by 30% when compared to conventional pavements with storm water drainage facilities. The FHWA reports the installation cost of PC to be 15–20% more than the conventional PCC. Currently, PCs are recommended for low volume roads, local streets, walkways and driveways. Research continues on the development of high strength PC that could be used on highways in the future.

10.3 PERMEABLE INTERLOCKING CONCRETE PAVEMENT (PICP)

Guidelines for design, construction and maintenance of PICP (Figure 10.9) have been developed by the ASCE (Smith and Hein, 2013). The maximum allowable traffic on PICP is 1 million (Equivalent Single Axle Load, ESAL, based on a standard 80kN axles). PICP has been used in urban alleys, parking lots and streets.

From top to bottom, the PICP consists of paver units with permeable jointing materials, over an open graded bedding course, an open graded base and/or an open graded subbase and the subgrade (Figure 10.10).

FIGURE 10.9 PICP in Berkeley, California. (With kind permission from Mr. David Hein, P.Eng.)

The concrete paver units are assembled in a manner so as to result in joints or openings. A 45 or 90° herringbone pattern has been recommended for PICP with vehicular traffic. The paver units have a maximum aspect (length to thickness) of 3 cm, and a minimum thickness of 8 cm. The jointing materials meet the gradation requirements of ASTM No. 8, 89 or 9. The open graded bedding material is typically 50 mm thick and conform to ASTM No. 8 stone or similar same-sized stone specification. The aggregate base layer consists of crushed stone from 2.5 cm to 1.3 cm in size, conforming to ASTM No. 57 specification. The subbase layer consists of materials ranging in size from 7.5 cm to 5 cm, conforming to ASTM No. 2, 3 or 4 specification. The subbase layer may not be provided for residential driveway or pedestrian path applications. A suitable geotextile layer is provided between the base/subbase and the subgrade to ensure protection from migration of fines and clogging of the open graded material.

In a full infiltration PICP the water fully infiltrates into the subgrade. In a partial infiltration PICP, part of the water infiltrates the subgrade, while the remaining is channeled through an underdrain in the subgrade. In the case of a no infiltration PICP, the water below the base/subbase is channeled solely through the underdrain, with an impermeable geomembrane between the base/subbase and the subgrade to prevent any water from escaping to the subgrade. The partial and low infiltration

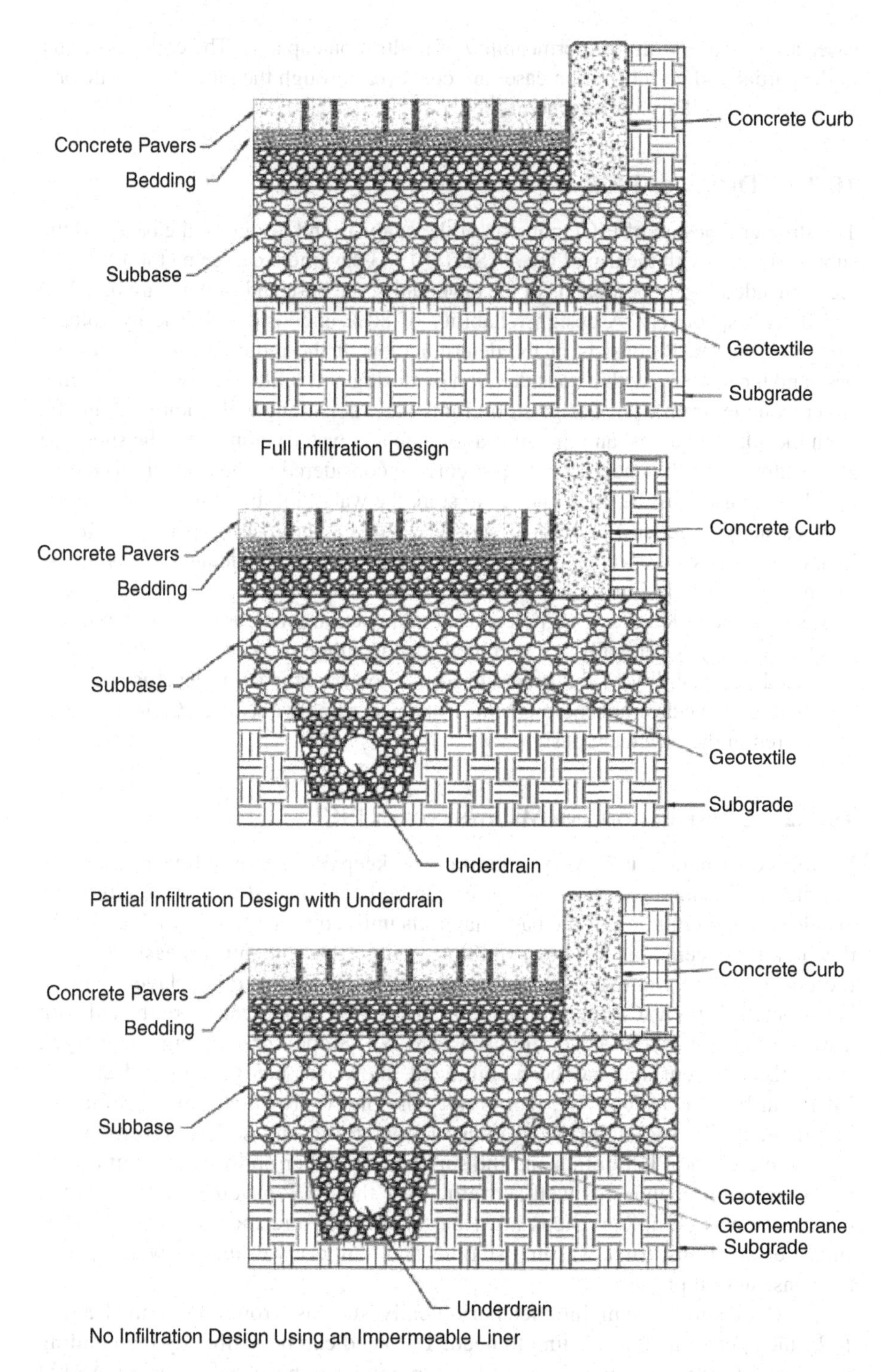

FIGURE 10.10 Different types of PICP. (With kind permission from Mr. David Hein, P.Eng.)

cases are for soils with low permeability of infiltration capacity. The collected water in the partial and no infiltration cases are conveyed through the pipes to a swale or a stream or an ancillary outlet system.

10.3.1 Design of PICP

The structural design of PICP consists of the analysis and design of the base and the subbase layers, with the use of the AASHTO 1993 pavement design method. Typical recommended layer coefficients of the paver units, the base and subbase are 0.3, 0.09 and 0.06, respectively. A design reliability of 80% is considered. The hydrologic design of the PICP consists of the determination of the required thickness of the base and the subbase with a consideration of storing and infiltrating (and/or draining water by underdrain). The design storm for the location along with additional runoffs from the adjacent areas, and the infiltration rate and water volume into the subgrade and outflow from the underdrain (if present) are considered in the analysis. The base and the subbase must be thick enough to store the water, infiltrate the water and also have capacity to receive additional surface water for an analysis period of 48–72 hours. The thicker section from the structural and the hydrologic analyses is finally recommended for design.

It should be noted that a high degree of compaction (such as 95% of maximum proctor density) is generally not specified for the subgrade – as that would lead to a reduced permeability of the subgrade soil. A lower compaction level is specified (depending on the desirable permeability), and accordingly a reduced CBR should be considered in the pavement design.

10.3.2 Construction and Maintenance of PICP

During construction it is very important to keep PICP areas free of dirt and sediments. Finished bases and subbases can be kept covered with geotextiles and an additional 5 cm thick of the base materials until construction traffic has slowed down and adjacent soil has been stabilized with erosion control measures. Then the cover can be removed and the final bedding and paver unit layers can be constructed. If the PICP layer is already constructed then it can be protected with a woven geotextile and a 5 cm thick ASTM No. 8 open graded aggregate layer. When the adjacent soil has been stabilized, the cover can be removed and the PICP can be swept clean. Or, if it is not possible to provide a cover, traffic can be allowed after the stoppage of construction traffic and soil stabilization and vacuuming of the PICP joints and the surface. The other option is to limit access to construction vehicles as well as traffic over the PICP pavement area until the construction is complete. Steps must be taken to ensure the use of silt fences, staged excavations, truck washing stations and temporary drainage swales during the construction phase.

The PICP should be maintained periodically, such as through vacuum cleaning 1–2 times per year or according to need. The need can be visible by the ponding of water after rainstorms. The infiltration rate can be checked using ASTM C1701, and rechecked after the vacuum, removal of dirty aggregates (generally to a depth of 1.3–2.5 cm), refilling of joints with clean aggregates and sweeping of the surface.

The conclusions from a seven-year study (Barrett and Diallo, 2017) on infiltration capacity of permeable pavements were that pervious concrete and PICP retained high IC (>1,000 cm/h), whereas porous asphalt lost about 86% of its original IC, down to a retained IC of 21 cm/h. The clogging was mostly found in the edge locations, and the interior locations showed much higher IC (≥68%).

10.4 SUSTAINABLE URBAN DRAINAGE SYSTEMS (SUDS)

Urbanization has a profound influence on drainage as a result of formation of more and more paved surfaces, an increase in the use of smooth straight channels and increase in runoff. In terms of discharge, effects can be summarized in two ways – a higher peak discharge and a decreased lag times. The net effect is felt in terms of decreased water quality, higher amounts of sedimentation and contaminants in water and an increase in the potential of flooding. Charlesworth et al. (2003) provides a comprehensive review of SUDs. Butler and Parkinson (1997) have laid out the desirable considerations for a sustainable urban drainage system which includes the avoidance of local or distant flooding, effective health barriers, avoidance of local or distant pollution, minimum utilization of natural resources and using long-term strategies which can be adapted to future requirements (some of which are unknown). The objective of a proper storm water drainage system is to reduce the peak flood flows, control pollution and improve the amenities (Argue and Pezzaniti, 1999). To fulfill these objectives sustainable drainage system should be able to reduce the quantity of runoff from a site (source control), slow the velocity of runoff to allow settlement filtration and infiltration (permeable filtration system), provide a passive treatment to the collected surface water before discharging onto land or water (end of pipe system). While conventional drainage systems can meet part of the above requirements, additional concepts and structures (sustainable) can be used to protect and enhance water quality and biodiversity, prevent flooding, protect the watercourse from pollutants due to accidental spillage and minimize utilization of natural resources and enable long-term strategies that are adaptable in the future. Best Management Practices (BMP) has been defined as a multidisciplinary approach in applying appropriate technology to preserve the natural environment, enhance the living standards and improve the quality of life (Braune and Wood, 1999).

10.4.1 Components of SUDS

SUDS includes both procedural (education and advice) as well as structural techniques. SUDS have been used in many countries, notably Sweden which has been a pioneer. The SUDS structures consists of physical barriers of pollution and flood waters, sink for contaminants and to enable natural flows to be maintained in the surface water course during dry weather. These structures consist of wetland (retention pond), permeable pavements, grass swales and dry ponds (detention), infiltrating trenches, filter drains and soakaways. SUDS have been extensively used in reducing non-point pollution (for example, in Scotland, the UK, and Florida,

Maryland and Wisconsin in the US). Retention ponds or wetlands are constructed with vegetation to maintain a permanent supply of water on the ground surface. They are very effective in reducing the potential of downstream flooding by retaining runoff from extreme events, and removing pollutants particularly those that remain attached to particulates through absorption and precipitation. The pollutant removal works this way – the flow velocity of the water decreases as it reaches the pond and this helps the particulates and their contaminants to settle out of the suspension. Over time the pond gets filled up with silt and this leads to a decrease in the efficiency of the water retaining capacity. To avoid this, regular dredging is required. A wetland is created by excavating land to a shallow depth, filling with earth, rock or gravel and then filling with water, and planting aquatic vegetation. The vegetation acts as a surface flow retention structures. They are low-cost structures in terms of both initial investment as well as operation. Wetlands improve water quality by removing large quantities of particulate and soluble contaminants, by systematically taking up dissolved constituents and entrapping particulate matters. Results show that wetlands can remove almost all bacteria and suspended solids and about 50% of phosphorous and nitrogen (Ellis, 1992). A secondary benefit provides habitat for wildlife and areas for recreation and education. Long-term maintenance required for the proper operation for wetlands include dredging, removal and replacement of plants. Wetlands can release contaminants previously locked up in sediments and in living tissue during die back in autumn and winter (Oberts and Osgood, 1991) and transform elements to bio available form (Mitsch and Gooselink, 2000). These should be taken into account when planning for the installation of wetlands. Porous pavements can be designed with integrated reservoir structures underneath that can work as effective in situ aerobic bioreactors. They have been reported to reduce petroleum contamination in the effluent by 97.6% of original oil applied on the surface (Pratt, 1999). Grass swales are shallow, grass lined channels for conveyance, storage, infiltration and treatment of storm water (IAWQ, 1996). Storm water runoff from adjacent structures flow in the swale for storage and infiltration or for conveyance to a different area. This delays storm water runoff peaks and reduces overall runoff volume through infiltration and evaporation. Grass swales work by removing particulate associated contaminants by sedimentation, filtration through grass lining and absorption into soil particles after infiltration. Swales are constructed with <5% shallow slopes and with well-draining soils, generally with a bottom width of about 1 m and a depth of 0.25 m to 2 m. Swales of length 30–60 m have been found to be able to retain 60–70% of solids and 30–40% of metals, hydrocarbons and bacteria. Potential problems and observations are as follows:

1. Sedimentation concentrations decrease along the length with maximum at the point where the water enters the grass strip.
2. Particles <6 um are unlikely to settle.
3. Sediment deposit is a slow process.

Dry ponds or detention ponds are depressions in the ground which may contain grass and can be constructed with concrete and can be used as storm water control and retention purpose. These get filled up with water during high rainfall and store the storm water. They also help allow pollutant sediment to settle out and infiltration if there is grass. There is a perforated pipe in dry ponds which permits the slow removal of the storm water. This way it decreases the peak flow and increases lag times. Dry ponds are not as effective as wet ponds for pollutant removal because of erosion of previously deposited silts when the pond fills up again (Benelmouffok and Yu, 1989).

10.4.2 Underground Structures and Drains

Underground structures are surface structures that allow the infiltration of water. Examples are infiltration trenches, inlets and curbs which can reduce storm water runoff and enable groundwater recharge (Fujita, 1994, Warnaars et al., 1999). South Australia (Argue and Pezzaniti, 1999). In an infiltration trench, a linear excavation is made and lined with filter fabric and back filled with stone. Grass is applied over the entire area and the runoff is diverted into the trench for infiltration and/or evaporation. Soakaways are constructed underground with precast concrete rings or dry walls and filled with stone (Butler and Davies, 2000). Metal and hydrocarbon concentrations have been found to be high in the top 1 cm of the soakaway surface and decrease rapidly with depth. They have been found to last for approximately 30 years.

Filter/French drains are located alongside highways together with a pipe that transmits water into the drainage system. They are much larger that soakaways with similar function. They are made up of a trench filled with gravel and wrapped in a membrane. The water flows through them and the top of the pipe drainage system (Butler and Davies, 2000). Water is stored, filtered and transported at a slow pace, and if sufficient water is lost by evaporation and infiltration there may not be a need for drainage system at all (SEPA, 2000).

10.4.3 Implementation

SUDS is an attractive option because (Urbonas and Stahre, 1993) it can help in the prevention of pollution at the source which is less costly than mitigating pollution further downstream. Pollution can be controlled by using both structural and non-structural/natural facilities and while onsite infiltration and percolation can reduce the volume of surface runoff, detention, retention and filtration and can also reduce pollutant conservation downstream. Delay in innovation and implementation of SUDS (Butler and Davies, 2000) can be caused by a lack of sufficient open space, concern about the level of maintenance, the lack of field data for comparison with conventional drainage system and concerns about legal adoption of schemes.

The followings steps have been suggested for successful implementation of SUDS (Ellis, 2001):

1. Incorporate the concept within the local drainage planning authorities and do not leave it to the environmental protection agencies.
2. Allocate appropriate funds for maintenance of the structures.

3. Assign responsibilities for these structures.
4. Educate people and students.
5. Conduct long-term monitoring for collecting data on performance and longevity.

10.4.4 Applicability of Specific SUDS Structures

Planning of SUDS is site specific, and some structures may not be applicable in certain situations; for example, from a space and safety point of view wetlands are not possible and desirable in urban areas, whereas porous pavements are. Also the efficiency of the different SUDS structures for removing certain pollutants should be considered to make a choice. Several structures can be used in combinations or stringed together to achieve a more holistic and integrated approach. An example is a combination of grass filter strips, wetlands and ponds, which have been proved to be effective in reducing storm water runoff (Rickard, 2002).

REFERENCES

Argue, J.R. and Pezzaniti, D. 1999. Catchment 'greening' using stormwater in Adelaide, South Australia. *Water Science and Technology*, 39 (2): 177–183.

Barrett, K.R. and Diallo, M. 2017. Long-term infiltration capacity of different types of permeable pavements. Final report for University Transportation Research Center, Region 2, grant number 49198-27-26 August 2017.

Benelmouffok, D.E. and Yu, S.L. 1989. Two dimensional numerical modelling of hydrodynamics and pollutant transport in a wet detention pond. *Water Science and Technology*, 21: 727–738.

Braune, M.J. and Wood, A. 1999. Best management practices applied to urban runoff quantity and quality control. *Water Science and Technology*, 39 (12): 117–121.

Butler, D. and Davies, J.W. 2000. *Urban Drainage*. London: E & FN Spon.

Butler, D. and Parkinson, J. 1997. Towards sustainable urban drainage. *Water Science and Technology*, 35: 53–63.

Chandrappa, A. and Biligiri, K. 2016. Pervious concrete as a sustainable pavement material – Research findings and future prospects: a state-of-the-art review. *Construction and Building Materials*, 111: 262–274.

Charlesworth, S.M., Harker, E. and Rickard, S. 2003. A review of sustainable drainage systems (SuDS): a soft option for hard drainage. *Geography*. 88 (2): 99–107.

Deo, O. and Neithalath, N. 2011. Compressive response of pervious concretes proportioned for desired porosities. *Construction and Building Materials*, 25: 4181–4189.

Ellis, J.B. 1992. 'Quality issues of source control', proceedings of CONFLO 92: integrated catchment planning and source control, Oxford.

Ellis, J.B. 2001. 'Managing urban drainage: policy and technical guidance needs', Centre for Environmental Research and Consultancy, Conference Proceedings 31 January, Coventry University, Coventry.

FHWA. 2015. Porous Asphalt Pavements with Stone Reservoirs. FHWA-HIF-15-009 April 2015.

Fu, T.F., Yeih, W., Chang, J.J. and Huang, R. 2014. The influence of aggregate size and binder material on the properties of pervious concrete. *Journal of Advances in Materials Science and Engineering*, 2014: 17.

Fujita, S. 1994. Infiltration structures in Tokyo. *Water Science and Technology*, 30 (1): 33–41.

Illgen, M. 2008. Infiltration and surface runoff processes on pavements: Physical phenomena and modelling. 11th International Conference on Urban Drainage, Edinburgh, Scotland, UK, 2008.

Illgen, M., Harting, K., Schmitt, T.G. and Welker, A. 2007. Runoff and infiltration characteristics of pavement structures – review of an extensive monitoring program. *Water Science and Technology*, 56 (10): 133–140.

International Association on Water Quality (IAWQ). 1996. Nature's Way – Designing for pollution prevention. https://wedc-knowledge.lboro.ac.uk/details.html?id=5083.

Jackson, N. 2003. *Design, Construction and Maintenance Guide for Porous Parking Lots*. National Asphalt Pavement Association, Information Series IS-131, October.

Kandhal, P.S. 2002. *Design, Construction, and Maintenance of Open-Graded Asphalt Friction Courses*. National Asphalt Pavement Association Information Series 115, May.

Kandhal, P.S. 2011. A Revolutionary Technique of Rainwater Harvesting Integrated into the Design of Buildings and Parking Lots. *Water Digest Magazine*, March–April 2011, New Delhi, India.

Kandhal, P.S. 2013. Role of Permeable Pavement in Groundwater Recharge. Presentation at the Rajasthan State Workshop on Water Conservation: Issues and Challenges. Held in Jaipur by the Centre for Science and Environment (CSE), 7 February 2013.

Kandhal, P.S. and Mishra, S. 2014. Design, Construction and Performance of Porous Pavement for Rainwater Harvesting. Indian Roads Congress. Indian Highways, March 2014.

Kuang X., Sansalone, J., Ying, G. and Ranieri, V. 2010. Pore-structure models of hydraulic conductivity for permeable pavement. *Journal of Hydrology*, 399: 148–157.

Mitsch, W.J. and Gosselink, J.G. 2000. *Wetlands (third edition)*. Chichester: Wiley.

Nguyen, D.H., Sebaibi, N. Boutouil, M., Leleyter, L. and Baurd, F. 2014. A modified method for the design of pervious concrete mix. *Construction and Building Materials*, 73: 271–282.

Oberts, G.L. and Osgood, P.A. 1991. Water quality effectiveness of a detention/wetland treatment system and its effect on an urban lake. *Environmental Management*, 15 (1): 131–138.

Pratt, C. 1999. Use of permeable reservoir pavement constructions for stormwater treatment and storage for re-use. *Water Science and Technology*, 39 (5): 145–151.

Rickard, S. 2002. The Management Issues of Sustainable Drainage Systems. Unpublished undergraduate dissertation. Coventry University.

Scottish Environment Protection Agency (SEPA). 2000. *Sustainable Urban Drainage Systems: An introduction*. Edinburgh: SEPA.

Smith, D.R and Hein, D.K. 2013. Development of a National ASCE Standard for Permeable Interlocking Concrete Pavement Second Conference on Green Streets, Highways, and Development November 3–6, 2013 | Austin, TX.

Thelen, E. and Howe, L.F. 1978. *Porous Pavement*. The Franklin Institute Research Laboratories.

Urbonas, B. and Stahre, R 1993. *Stormwater: Best Management Practices and Detention for Water Quality, Drainage and CSO Management*. New Jersey: Prentice Hall.

Warnaars, E., Veildt Larsen, A., Jacobsen, R. and Steen Mikkelsen, R. 1999. Hydrologic behaviour of stormwater infiltration trenches in a central urban area during 2/3 years of operation. *Water Science and Technology*, 39 (2): 217–224.

Yahia, A. and Kabagire, D. 2014. New approach to proportion pervious concrete. *Construction and Building Materials*, 62 (March); 38–46.

Index